LABORATORY DECONTAMINATION AND DESTRUCTION OF CARCINOGENS IN LABORATORY WASTES: SOME MYCOTOXINS

INTERNATIONAL AGENCY FOR RESEARCH ON CANCER

The International Agency for Research on Cancer (IARC) was established in 1965 by the World Health Assembly, as an independently financed organization within the framework of the World Health Organization. The headquarters of the Agency are at Lyon, France.

The Agency conducts a programme of research concentrating particularly on the epidemiology of cancer and the study of potential carcinogens in the human environment. Its field studies are supplemented by biological and chemical research carried out in the Agency's laboratories in Lyon and, through collaborative research agreements, in national research institutions in many countries. The Agency also conducts a programme for the education and training of personnel for cancer research.

The publications of the Agency are intended to contribute to the dissemination of authoritative information on different aspects of cancer research. A complete list is printed at the back of this book.

WORLD HEALTH ORGANIZATION

INTERNATIONAL AGENCY FOR RESEARCH ON CANCER

MINISTRY OF THE ENVIRONMENT FRANCE

Laboratory Decontamination and Destruction of Carcinogens in Laboratory Wastes : Some Mycotoxins

Editors

M. Castegnaro, J. Barek, J.-M. Frémy, M. Lafontaine, M. Miraglia, E.B. Sansone & G.M. Telling

IARC Scientific Publications No. 113

International Agency for Research on Cancer
Lyon, France
1991

Published by the International Agency for Research on Cancer,
150 cours Albert Thomas, 69372 Lyon Cedex 08, France

Distributed by Oxford University Press, Walton Street, Oxford OX2 6DP, UK

Distributed in the USA by Oxford University Press, New York

ISBN 92 832 2113 3
ISSN 0300-5085

Printed in France

Contents

Foreword

There is a widespread interest in the field of mycotoxins and, while a large part of the work has been devoted to aflatoxins, there is increasing interest in other mycotoxins, especially ochratoxin A, which was recently demonstrated to be carcinogenic to two animal species.

In its programme for the development of degradation techniques for chemical carcinogens, IARC has decided to develop methods for four commonly investigated mycotoxins: citrinin, ochratoxin A, patulin and sterigmatocystin. In view of recent findings concerning the mutagenic potency of Mn^{2+} generated by potassium permanganate under acid conditions, the procedure for the degradation of mycotoxins with potassium permanganate under alkaline conditions, that does not generate mutagenic species, has been extended to aflatoxins.

For this programme IARC is pleased to acknowledge the sustained effort of the Division of Safety of the National Institutes of Health of the USA, and that of the French Ministry of the Environment (contract SRETIE / MÈRE / 88201). The Commission of Food Chemistry of the IUPAC included this project in its programme (Project No. 61/89), and several members participated in the development and/or validation work.

The Agency very much appreciates this support which, it is hoped, will contribute to the implementation of good laboratory practice in research institutes.

L. Tomatis, M.D.
Director, IARC

Mycotoxins considered

The following mycotoxins are considered in this volume. The methods described for the destruction of specific compounds may be applicable to others from the same group. However, when dealing with other compounds, the efficiency of the methods should first be verified.

Compound considered	Chemical Abstracts Services Registry Number	Abbreviation
Citrinin	518-75-2	
Ochratoxin A	303-47-9	OA
Patulin	149-29-1	
Sterigmatocystin	10048-13-2	
Aflatoxin B_1	1162-65-8	AFB_1
Aflatoxin B_2	7220-81-7	AFB_2
Aflatoxin G_1	1165-39-5	AFG_1
Aflatoxin G_2	7241-98-7	AFG_2

Introduction

CARCINOGENICITY

The carcinogenicity of sterigmatocystin, citrinin, ochratoxin A and patulin has been evaluated by working groups of experts (International Agency for Research on Cancer, 1976, 1983, 1986, 1987). Sterigmatocystin has been classified in group IIB, i.e., the agent is possibly carcinogenic to humans. The other three compounds were classified in group III, i.e. the agent is not classifiable as to its carcinogenicity to humans.

However, in view of recently published data concerning the carcinogenic potency of OA to a second species of animal, the rat (Boorman, 1988), and its possible link to Balkan endemic nephropathy and the associated urothelial cancer (Castegnaro *et al.*, 1990), this compound has been placed high on a prioriry list for reevaluation by a working group of experts. No recent data seem to warrant reevaluation of citrinin and patulin.

ANALYSIS

A number of methods have been investigated and collaboratively tested by the Association of Official Analytical Chemists (Williams, 1984), for the analysis of OA in barley (26-111 to 26-118) and in green coffee (26-119 to 26-125), patulin in apple juice (26-126 to 26-131) and of sterigmatocystin in barley and wheat (26-132 to 26-138). They can easily be adapted to the analysis of residues in wastes after treatment. All these methods make use of thin-layer chromatography (TLC), but with the development of high-pressure liquid chromatography (HPLC), several techniques have been published which make use of UV spectroscopy or spectrofluorimetry for detection. Some of them have been published in Stoloff *et al.* (1982) and a selection is presented in Table 1. Patulin (Ogawa *et al.*, 1984; Skierska & Martinek, 1984; Ehman & Gaucher, 1977) and sterigmatocystin (Suzuki *et al.*, 1976; Salhab *et al.*,

1976) have also been analysed by gas-liquid chromatography, this latter technique having been coupled to single and multiple ion monitoring mass spectrometry for the analysis of patulin (Price, 1979). Immunochemical techniques have also been proposed for the analysis of OA in food, feed and blood (Candlish *et al.*, 1988; Kawamura *et al.*, 1989; Chu *et al.*, 1976; Morgan *et al.*, 1983, 1986; Rousseau *et al.*, 1987). The various immunochemical techniques for analysis of mycotoxins have been reviewed by Morgan (1989).

A semi-quantitative TLC method for the determination of citrinin in corn and barley has been described by Wilson in Stoloff *et al.* (1982). Other methods by Chalam and Stahr (1979), Gimeno and Martins (1983) and Gimeno (1984) proposed the use of TLC, and that of HPLC with spectrofluorimetric detection (Lepom, 1986b; Marti *et al.*, 1978) or UV detection (Phillips *et al.*, 1980). It can also be analysed directly on the extract by fluorimetry (Tranthan & Wilson, 1984).

Table 1. Selection of methods using HPLC for analysis of mycotoxins

Compound analysed	Detection system	Reference
Patulin	UV (254 nm)	Forbito & Babsky (1985)
	UV (280 nm)	Hunt *et al.* (1978)
Sterigmatocystin	UV (245 nm)	Hurst *et al.* (1987)
	UV (325 nm)	Lepom (1986a)
	UV (246 nm)	Schmidt *et al.* (1981)
Ochratoxin A	Fluorescence (excit.: 330 nm, em.: 460 nm)	Cohen & Lapointe (1986)
	Fluorescence (excit.: 330 nm, em.: 418 nm)	Frohlich *et al.* (1988)
Citrinin + (other)	UV (254 nm)	Frisvad (1987)
Citrinin, ochratoxin A, patulin & sterigmatocystin	UV (254 nm)	Bicking & Kniseley (1980)

Methods of degradation

Five methods have been tested for their capacity to degrade ochratoxin A (OA) into non-mutagenic residues : ammoniation, heat treatment, oxidation by potassium permanganate (0.3 mol/L) in sulfuric acid (3 mol/L) and in sodium hydroxide (2 mol/L) and oxidation by sodium hypochlorite. Ammoniation was discarded as OA was reformed after neutralization, whether treatment was at room temperature or in boiling solution. Heat treatment in aqueous solution had no effect on the degradation of OA. Complete disappearance of the OA molecule was achieved by oxidation with 0.3 mol/L potassium permanganate in 3 mol/L sulfuric acid, but the residues obtained were mutagenic without activation in *Salmonella typhimurium* TA97, TA98, TA100 and TA102 strains. This mutagenic activity has been attributed to the formation of Mn^{2+}. Only treatment with sodium hypochlorite or alkaline potassium permanganate gave satisfactory results for both efficiency of degradation and absence of mutagenicity of the residues. These are described as Methods 1 and 5, respectively.

Four methods for the degradation of citrinin have been evaluated: treatment with ammonia, oxidation by sodium hypochlorite and oxidation by 0.3 mol/L potassium permanganate in sulfuric acid (3 mol/L), and in alkaline conditions (2 mol/L). All these methods led to complete removal of the molecule. The third technique (oxidation by potassium permanganate in sulfuric acid) produced residues which were mutagenic without activation to *Salmonella typhimurium* strains TA100 and TA102. This mutagenic activity has been attributed to the formation of Mn^{2+}. The three other techniques gave satisfactory results and are described as Methods 2, 1 and 5, respectively.

Four methods for the destruction of sterigmatocystin have been evaluated: treatment with 6 mol/L sulfuric acid, oxidation by 0.3 mol/L potassium permanganate in 3 mol/L sulfuric acid and in sodium hydroxide (2 mol/L) and oxidation by sodium hypochlorite. All four methods led to complete loss of the original molecule, but mutagenic residues were obtained by the first two methods. Only treatment with sodium hypochlorite and treatment with potassium permanganate in alkaline conditions proved satisfactory. These are described as Methods 3 and 5, respectively.

Three techniques have been tested for the degradation of patulin: treatment with ammonia and treatment with potassium permanganate in acidic (3 mol/L sulfuric acid) or alkaline (2 mol/L sodium hydroxide) conditions. The technique using potassium permanganate in sulfuric acid produced residues which were mutagenic without activation to *Salmonella typhimurium* strains TA100 and TA102. This mutagenic activity has been attributed to the formation of Mn^{2+}. The technique using ammonia is described as Methods 4 and 6, and that using potassium permanganate in alkaline conditions as Method 5.

The criterion for significant mutagenicity was set at twice the level of the control value.

One of the methods described (Method 5), using potassium permanganate in alkaline conditions, is suitable for degradation of aflatoxins. Other methods for degradation of aflatoxins have been described in an earlier volume of this series (Castegnaro *et al.*, 1980).

Collaborating organizations

The methods in this document have been tested by the following collaborating laboratories and their description benefited from the work of the group.

Dr J. BAREK, Dr J. MATĚJKA and Dr J. ZIMA
Department of Analytical Chemistry
Charles University
128 40 Prague 2, Czechoslovakia

Dr J.-M. FREMY and Dr E. GLEIZES
Ministère de l'Agriculture — CNEVA
Laboratoire Central d'Hygiène Alimentaire
43 rue de Dantzig
75015 Paris, France

Dr M. LAFONTAINE and Dr Y. MORELE
Institut National de Recherche et de Sécurité
Avenue de Bourgogne
54500 Vandœuvre, France

Dr M. MIRAGLIA and Dr C. BRERA
Istituto Superiore de la Sanità
Laboratorio Alimenti
Viale Regina Elena 299
00161 Rome, Italy

Dr G.M. TELLING and Dr J. MILLER
Unilever Research Colworth Laboratory
Colworth House
Sharnbrook, Beds, MK44 1LQ, UK

Dr E.B. SANSONE and Dr G. LUNN
NCI-Frederick Cancer Research and Development Center
P.O. Box B
Frederick, MD 21702, USA

International Agency for Research on Cancer
150 cours Albert Thomas
69372 Lyon Cedex 08, France

Methods Index

METHODS RECOMMENDED FOR SPECIFIC WASTE CATEGORIES

Waste category	Recommended destruction method (in order of preference)				
	OA	Citrinin	Sterigmato-cystin	Patulin	Aflatoxins[a]
Solid compound	1, 5	1, 5, 2	3, 5	5, 6	5
Aqueous solution	1, 5	1, 5, 2	3, 5	5, 6	5
Solutions in volatile organic solvents	1, 5	1, 5, 2	3, 5	5, 6	5
Solutions in DMF or DMSO	1	1	3		
Solutions in non-volatile organic solvents miscible with water		2			
Glassware or protective clothing	1, 5	1, 5, 2	3, 5	5, 6	5
Litter				4	
TLC plates	1	1	3		
Spills	1, 5	1, 5	3, 5	5	5

[a] See Castegnaro *et al.* (1980) for other destruction methods for aflatoxins.

Method 1

Destruction of ochratoxin A and citrinin using sodium hypochlorite

1. SCOPE AND FIELD OF APPLICATION

The method specifies a procedure for the destruction of ochratoxin A (OA) and citrinin in various laboratory wastes: solid compound (7.1); aqueous solutions (7.2); solutions in volatile organic solvents (7.3); solutions in dimethylformamide (DMF) or dimethylsulfoxide (DMSO) (7.4); glassware and protective clothing (7.5); spills (7.6) and TLC plates (7.7).

The method has been collaboratively tested using 400 μg of OA in 500 μL ethanol and 400 μg solid citrinin. The method affords better than 99% degradation for OA and 99.5% degradation for citrinin.

The residues obtained by this method for degradation of OA and citrinin have been tested for mutagenic activity using *Salmonella typhimurium* strains TA97, TA98, TA100 and TA102 with and without metabolic activation. No mutagenic activity was detected.

2. REFERENCES

2.1 To the method: Stoloff, L. & Trager, W. (1965) Recommended decontamination procedures for aflatoxins. *J. Assoc. Off. Anal. Chem.*, **48**, 681-682

2.2 To handling: Castegnaro, M., Van Egmond, H.P., Paulsch, W.E. & Michelon, J. (1982) Limitation in protection afforded by gloves in laboratory handling of aflatoxins. *J. Assoc. Off. Anal. Chem.*, **65**, 1520-1523

3. PRINCIPLE

Solid citrinin and solutions of OA or citrinin in a small volume of an alcohol are completely degraded by treatment with an excess of sodium hypochlorite solution 5°Cl.

4. HAZARDS

4.1 *Ochratoxin and citrinin*

Ochratoxin A is teratogenic and carcinogenic. Citrinin is nephrotoxic and there is limited evidence of its carcinogenicity in experimental animals. It is recommended that handling of these compounds should always be carried out in ventilated enclosures. Gloves and other protective clothing must be worn for all operations involving handling of these compounds or their solutions.

Moreover, it has been demonstrated that solutions of aflatoxins in chloroform can diffuse through latex and vinyl gloves. Should gloves come into contact with a mycotoxin solution, they should therefore be changed as quickly as possible, to reduce the risk of contact of the mycotoxin with the skin.

To avoid problems of dispersion due to electrostatic effects, OA or citrinin in the powder form should be handled using cotton gloves.

4.2 *Other hazards*

Sodium hypochlorite is a strong oxidizing agent; care must be taken not to mix it with concentrated reducing or other oxidizable substances. In the case of skin contact with corrosive chemicals, wash the skin under flowing water for at least 15 min.

5. REAGENTS

5.1 *For destruction*

Water	Deionized
Ethanol	Technical product
Methanol	Technical product
Dichloromethane	Technical product
Sodium bicarbonate	Technical product
Sodium hypochlorite solution (5°Cl)	Technical product
Sodium hydroxide	Technical product
Sodium sulfate (anhydrous)	Technical product

NOTE 1: *Hypochlorite solution*

It must be remembered that solutions of sodium hypochlorite tend to deteriorate. It is therefore essential to check their available chlorine content before use.

Note that strength of sodium hypochlorite solutions may be given as weight/weight or weight/volume. This is an additional reason for estimating the concentration of available chlorine.

NOTE 2: *Definition*

Percent (%) available chlorine = mass of chlorine, in grams, liberated by acidifying 100 g of sodium hypochlorite solution. The available chlorine may also be expressed as °Cl, which corresponds to the volume of chlorine, in litres, liberated by one litre (liquid) or one kilogram (solid) of commercial bleach when treated with hydrochloric acid, e.g., a 1 mol/L solution of hypochlorite corresponds to 22.4°Cl.

NOTE 3: *Determination of hypochlorite strength*

The sodium hypochlorite solution used for this determination should be adjusted to contain not less than 25 g and not more than 30 g of available chlorine per litre, by diluting the purchased solution of approximately known strength. Assay: Pipette 10.0 mL of sodium hypochlorite solution into a 100 mL volumetric flask and fill to the mark with distilled water. Pipette 10.0 mL of the resulting solution into a conical flask which already contains 50 mL of distilled water, 1 g of potassium iodide and 12.5 mL of acetic acid (2 mol/L), rinse and titrate with (0.05 mol/L) sodium thiosulfate, using starch as indicator. 1 mL sodium thiosulfate (0.05 mol/L) corresponds to 3.545 mg available chlorine.

5.2 *For analysis*

Ethyl acetate	Analytical grade
Toluene	Analytical grade
Dichloromethane	Analytical grade
Methanol	HPLC grade
Acetonitrile	HPLC grade
Water	Deionized
Sodium acetate	Analytical grade
Sodium sulfate	Analytical grade, anhydrous
Acetic acid	Glacial—analytical grade
Phosphoric acid	Analytical grade
Hydrochloric acid	12 mol/L and 1 mol/L—analytical grade
Formic acid	Analytical grade
Oxalic acid	Analytical grade

6. APPARATUS

Usual laboratory equipment and the following items: HPLC system equipped with a UV spectrophotometric or a spectrofluorimetric detector

TLC plates	Silica gel 60 without fluorescent indicator, treated for 1.5 min with 10% oxalic acid in methanol and dried overnight at room temperature
TLC developing chamber	
UV lamp	Long-wave
Ultrasound bath	
Rotary evaporator	

7. PROCEDURE

Solid citrinin (400 µg) or a solution of 400 µg OA or citrinin in 400 µL of ethanol (OA) or 800 µL of methanol (citrinin) is completely degraded by treatment with 10 mL of a 5°Cl hypochlorite solution for 30 min. Since other compounds in the waste may also react with sodium hypochlorite, it is recommended that the efficiency of the degradation be checked using the analytical procedure described in Section 8 or any other sensitive analytical procedure.

7.1 *Solid compound*

7.1.1 Estimate the amount to be degraded and thus the amount of hypochlorite solution to be used.

7.1.2 For each 400 µg OA or citrinin add 400 µL of ethanol (OA) or 800 µL of methanol (citrinin).

7.1.3 Add at least twice the estimated amount of hypochlorite required (i.e. 20 mL per 400 µg product).

7.1.4 Sonicate to improve solubilization and allow to react for at least 30 min.

7.1.5 Check for completeness of degradation.

7.1.6 Dilute and discard.

7.2 *Aqueous solutions*

7.2.1 Make neutral or alkaline, if necessary.

7.2.2 Proceed as in 7.1.1 to 7.1.6.

7.3 *Solutions in volatile organic solvents*

7.3.1 Evaporate solvent using a rotary evaporator under reduced pressure.

7.3.2 Estimate the amount of OA or citrinin to be degraded and add enough ethanol (OA) or methanol (citrinin) to wet the glass.

7.3.3 Proceed as in 7.1.2 to 7.1.6.

7.4 *Solutions in DMSO or DMF*

7.4.1 For each volume of solution add 2 volumes of water.

7.4.2 Extract three times with equal volumes of dichloromethane. Pool dichloromethane extracts and dry over anhydrous sodium sulfate. Filter off the sodium sulfate, washing with another volume of dichloromethane.

7.4.3 Evaporate solvent using a rotary evaporator under reduced pressure making sure that no solvent remains (if necessary raise temperature of the bath and vacuum).

7.4.4 Proceed as in 7.3.2. to 7.3.3.

7.5 *Glassware or protective clothing*

7.5.1 Add enough ethanol or methanol to wet the glass or protective clothing.

7.5.2 Immerse glassware or protective clothing in a 5°Cl sodium hypochlorite solution and leave in contact for 30 min.

7.5.3 Dilute decontaminating solution and discard.

7.6 *Spills*

7.6.1 Isolate the area and put on suitable protective clothing.

7.6.2 Collect liquid on an absorbing cloth or solid with a cloth wetted with a sodium bicarbonate solution.

7.6.3 Immerse cloth in a 5°Cl sodium hypochlorite solution bath.

7.6.4 Rinse area with a cloth wetted with a sodium bicarbonate solution and immerse cloth as in 7.6.3.

7.6.5 Cover the area with a 5°Cl sodium hypochlorite solution.

7.6.6 Allow to react for at least 30 min.

7.6.7 Absorb solution on absorbent and discard.

7.6.8 To check for completeness of degradation, wipe the surface with an absorbent material moistened with methanol and analyse the wipe.

7.7 *Thin-layer chromatography plates*

Spray the plates with a solution of hypochlorite (5°Cl) and allow to react for 30 min. The efficiency of the degradation can be checked by scraping the plate and eluting with a suitable solvent.

8. ANALYSIS FOR COMPLETENESS OF DEGRADATION

8.1 *HPLC conditions for ochratoxin A and citrinin*

8.1.1. Take an aliquot from 7.1.5, 7.2.2, 7.3.3, 7.4.4 or 7.5.2 and acidify slowly to about pH 3 – 4 using concentrated (12 mol/L) HCl under an efficient fume hood.

8.1.2 Complete removal of chlorine under the hood by bubbling nitrogen through the solution.

8.1.3 Analyse by HPLC using, for example, the following conditions or any other suitable system.
Column: 25 cm × 3.6 mm i.d. Partisil ODS-2 10 μm
Precolumn: 65 mm × 3.6 mm i.d. filled with CO: Pell ODS 30 – 38 μm
Mobile phase: Isocratic system. Acetonitrile: 0.25 mol/L aqueous H_3PO_4 (75:25)
Flow rate: 1.5 mL/min
Detector: For OA spectrofluorimetry, excitation 340 nm, emission 465 nm
For citrinin: UV 330 nm, spectrofluorimetry excitation 330 nm, emission 480 nm

8.2 *TLC conditions for citrinin*

8.2.1 Take an aliquot from 7.1.5, 7.2.2, 7.3.3, 7.4.4 or 7.5.2 and acidify slowly to about pH 3 – 4 using concentrated (12 M) HCl under an efficient fume hood.

8.2.2 Extract three times with an equal volume of dichloromethane and pool dichloromethane extracts.

8.2.3 Dry extract over anhydrous sodium sulfate and filter off the sodium sulfate.

8.2.4 Evaporate to near dryness using a rotary evaporator under reduced pressure, then to dryness under a gentle stream of nitrogen.

8.2.5 Take up residue in a minimal volume of dichloromethane.

8.2.6 Apply all to TLC plates prepared as describe in Section 6. Also apply a series of standards.

8.2.7 Develop plates in toluene:ethyl acetate:dichloromethane:formic acid (90:45:50:5).

8.2.8 View chromatogram under long-wave UV light.

9. SCHEMATIC REPRESENTATION OF THE PROCEDURE

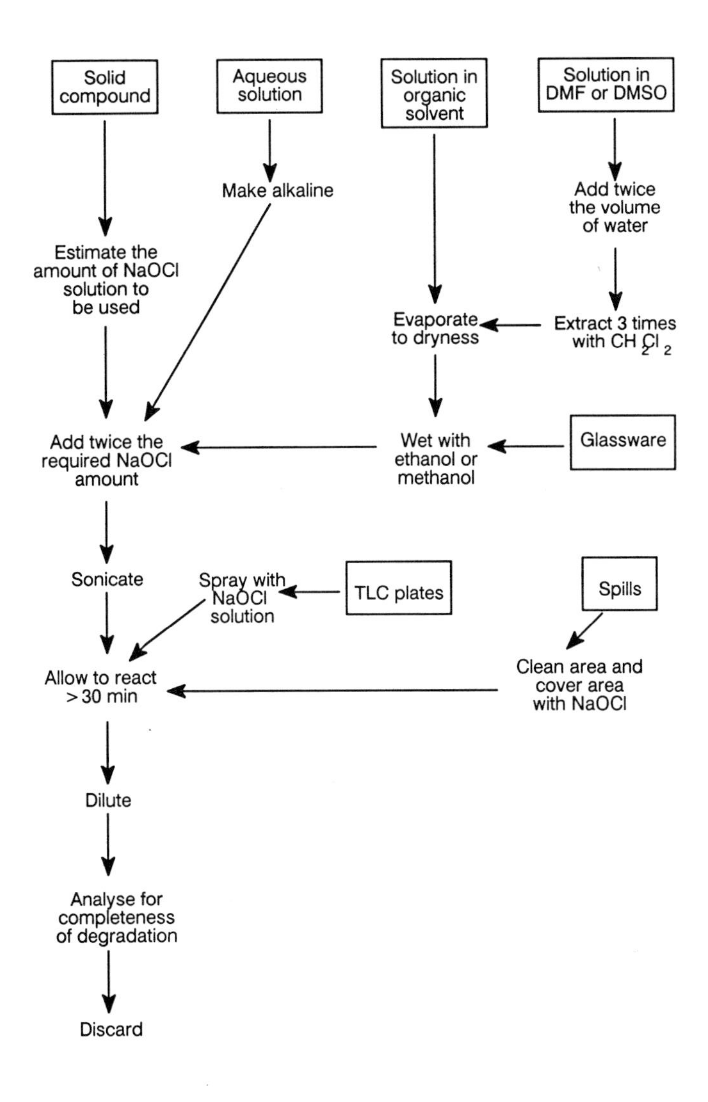

10. ORIGIN OF METHOD

M. Castegnaro and J. Michelon
IARC
150, cours Albert-Thomas
69372 Lyon Cedex 08, France

Contact point: M. Castegnaro

M. de Méo and M. Laget
Laboratoire de Microbiologie,
Faculté de Pharmacie
27, bd Jean-Moulin
13385 Marseille Cedex 05
France

Method 2

Degradation of citrinin using ammoniation

1. SCOPE AND FIELD OF APPLICATION

This method specifies a procedure for the destruction of citrinin in the following wastes: solid compounds (7.1); aqueous solutions (7.2); solutions in volatile organic solvents (7.3); solutions in organic solvents miscible with water (7.4) and glassware (7.5).

The method has been collaboratively tested using 200 μg of citrinin. In this case it affords better than 99.5% degradation.

The residues obtained by this method for degradation of citrinin have been tested for mutagenic activity using *Salmonella typhimurium* strains TA97, TA98, TA100 and TA102 with and without metabolic activation. No mutagenic activity was detected.

2. REFERENCES

2.1 To the method: Lee, L.S. & Cucullu, A.F. (1978) Conversion of aflatoxin B_1 to aflatoxin D_1 in ammoniated peanut and cottonseed meals. *J. Agric. Food Chem.*, **26**, 881-884

2.2 To handling: Castegnaro, M., Van Egmond, H.P., Paulsch, W.E. & Michelon, J. (1982) Limitation in protection afforded by gloves in laboratory handling of aflatoxins. *J. Assoc. Off. Anal. Chem.*, **65**, 1520-1523

3. PRINCIPLE

Destruction is effected by treatment with ammonia at 100°C in a tightly closed metal container such a bomb calorimeter.

4. HAZARDS

4.1 *From citrinin*

Citrinin is nephrotoxic and there is limited evidence of its carcinogenicity in experimental animals. It is recommended that handling of citrinin should always be carried out in a ventilated enclosure. Gloves and other protective clothing must be worn for all operations involving handling of this compound or its solutions.

Moreover, it has been demonstrated that solutions of aflatoxins in chloroform can diffuse through latex and vinyl gloves. Should gloves come into contact with a mycotoxin solution, they should therefore be changed as quickly as possible, to reduce the risk of contact of the mycotoxin with the skin.

To avoid problems of dispersion due to electrostatic effects, citrinin in powder form should be handled using cotton gloves.

4.2 *Others*

Heating an ammonia solution in a closed vessel leads to a build-up in pressure. This must be considered when selecting a suitable vessel. Ammonia solution is corrosive and gives off toxic ammonia gas.

5. REAGENTS

5.1 *For decontamination*

Ammonia solution	10 %, 20 % (w/w)
Methanol	Technical grade

5.2 *For analysis*

Hydrochloric acid	Analytical grade
Hydrochloric acid	1 mol/L, aqueous
Formic acid	Analytical grade
Sodium sulfate	Analytical grade, anhydrous
Dichloromethane	Analytical grade
Acetonitrile	HPLC grade
Toluene	Analytical grade
Ethyl acetate	Analytical grade
Oxalic acid	Analytical grade
Phosphoric acid	Analytical grade

6. APPARATUS

Usual laboratory equipment and the following items:

Tightly closed metal container such as a bomb calorimeter

HPLC system equipped with a UV spectrophotometric or spectrofluorimetric detector.

TLC plates	Silica gel 60 without fluorescent indicator. Treated for 1.5 min with 10% oxalic acid in methanol, dried overnight at room temperature
TLC developing chamber	
Rotary evaporator	
UV lamp	Long-wave
Ultrasound bath	

7. PROCEDURE

Four hundred micrograms of citrinin are degraded by treatment overnight with 10 mL 10% ammonia at 100°C.

7.1 *Solid compound*

7.1.1 For 400 micrograms of citrinin, add 10 mL of 10% ammonia solution. *Do not close glass flask tightly.*

7.1.2 Sonicate about 1 min to improve solubilization.

7.1.3 Place in a bomb calorimeter, close tightly and put it in an oven at 100°C.

7.1.4 Allow to react overnight.

7.1.5 Allow to come to room temperature.

7.1.6 Test for efficiency of degradation (see section 8).

7.1.7 Dilute and discard.

7.2 *Aqueous solutions*

7.2.1 Add and equivalent volume of a 20% ammonia solution.

7.2.2 Proceed as in 7.1.3 to 7.1.7.

7.3 *Solutions in volatile organic solvents*

7.3.1 Estimate the amount of citrinin to be degraded.

7.3.2 Evaporate solvent using a rotary evaporator under reduced pressure.

7.3.3 Proceed as in 7.1.1 to 7.1.7.

7.4 *Solutions in organic solvents miscible with water*

7.4.1 Estimate the amount of citrinin to be degraded.

7.4.2 Add at least the same volume of a 20% ammonia solution, or more if required.

7.4.3 Proceed as in 7.1.3 to 7.1.7.

7.5 *Glassware*

7.5.1 Rinse five times with enough methanol to wet the glass.

7.5.2 Treat rinse as in 7.4.

8. ANALYSIS FOR COMPLETENESS OF DEGRADATION

8.1 *TLC conditions*

8.1.1 Take an aliquot from 7.1.7, 7.2.2, 7.3.2, 7.4.3 or 7.5.2 and acidify to about pH 3 using HCl.

8.1.2 Extract three times with an equal volume of dichloromethane and pool dichloromethane extracts.

8.1.3 Dry extract over sodium sulfate and filter off the sodium sulfate.

8.1.4 Evaporate filtrate to near dryness using a rotary evaporator under reduced pressure, then to dryness under a gentle stream of nitrogen.

8.1.5 Take up residue in the minimum amount of dichloromethane.

8.1.6 Apply all solution to TLC plates prepared as described in Section 6. Also apply a series of standards.

8.1.7 Develop plates in toluene:ethyl acetate:dichloromethane:formic acid (90:45:50:5).

8.1.8 View chromatogram under long-wave UV light.

8.2 *HPLC conditions*

8.2.1 Take an aliquot from 7.1.7, 7.2.2, 7.3.2, 7.4.3 or 7.5.2 and acidify to about pH 3 using HCl.

8.2.2 Extract three times with an equal volume of dichloromethane and pool dichloromethane extracts.

8.2.3 Dry extract over sodium sulfate and filter off the sodium sulfate.

8.2.4 Evaporate filtrate to near dryness using a rotary evaporator under reduced pressure, then to dryness under a gentle stream of nitrogen.

8.2.5 Take up residue in acetonitrile.

8.2.6 Analyse by HPLC using, for example, the following conditions or any other suitable system.
Column: 25 cm × 3.6 mm i.d. Partisil ODS-2 10 μm
Precolumn: 65 mm × 3.6 mm i.d. filled with CO:Pell ODS 30-38 μm
Mobile phase: Isocratic system. Acetonitrile: 0.25 mol/L aqueous H_3PO_4 (75:25)
Flow rate: 1.5 mL/min
Detector: For citrinin: UV 330 nm, spectrofluorimetry excitation 330 nm, emission 480 nm.

9. SCHEMATIC REPRESENTATION OF THE PROCEDURE

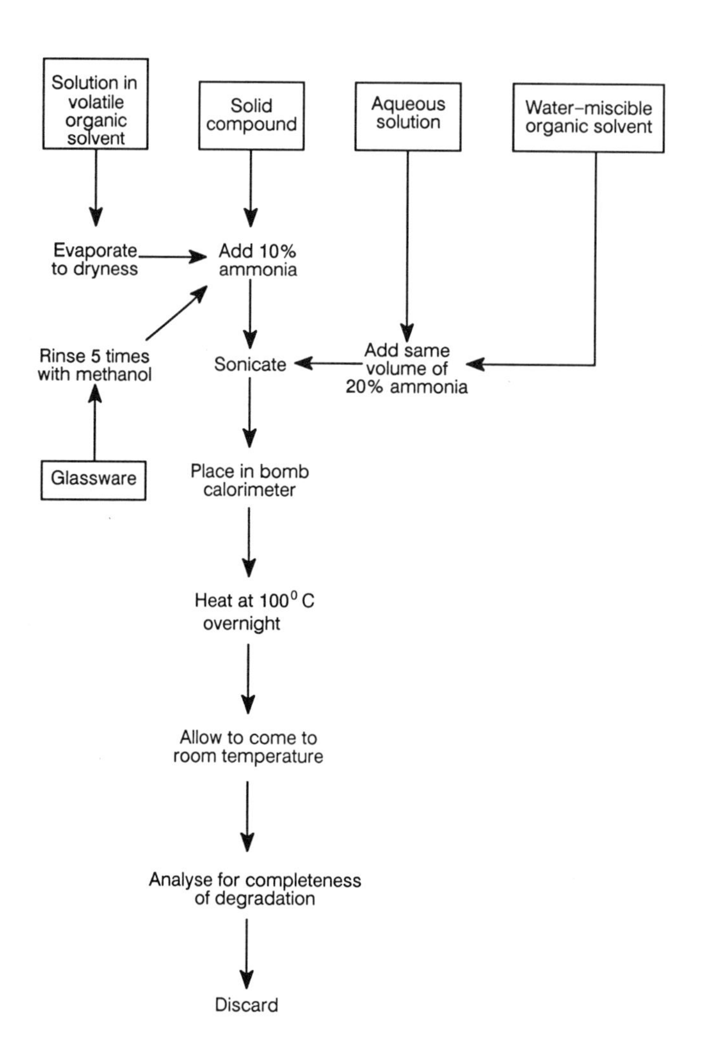

10. ORIGIN OF METHOD

M. Castegnaro and J. Michelon
IARC
150, cours Albert-Thomas
69372 Lyon Cedex 08, France

Contact point: M. Castegnaro

M. De Meo and M. Laget
Laboratoire de Microbiologie
Faculté de pharmacie
27, bd Jean-Moulin
13385 Marseille Cedex 05, France

Method 3

Destruction of sterigmatocystin using sodium hypochlorite

1. SCOPE AND FIELD OF APPLICATION

This method specifies a procedure for the destruction of sterigmatocystin in the following laboratory wastes: solid compound (7.1); aqueous solutions (7.2); solutions in volatile organic solvents (7.3); solutions in dimethylsulfoxide (DMSO) or dimethylformamide (DMF) (7.4); glassware and protective clothing (7.5); spills (7.6) and TLC plates (7.7).

The method has been collaboratively tested using 200 μg of sterigmatocystin. In this case it affords better than 99.8% degradation.

The residues obtained by this method for degradation of sterigmatocystin have been tested for mutagenic activity using *Salmonella typhimurium* strains TA97, TA98, TA100 and TA102 with and without metabolic activation. No mutagenic activity was detected.

2. REFERENCES

2.1 To the method: Stoloff, L. & Trager, W. (1965) Recommended decontamination procedures for aflatoxins. *J. Assoc. Off. Anal. Chem.*, **48**, 681-682

Castegnaro, M., Friesen, M., Michelon, J. & Walker, E.A. (1981) Problems related to the use of sodium hypochlorite in the detoxification of aflatoxin B_1. *Am. Ind. Hyg. Assoc. J.*, **42**, 398-401

2.2 To handling: Castegnaro, M., Van Egmond, H.P., Paulsch, W.E. & Michelon, J. (1982) Limitation in protection afforded by gloves in laboratory handling of aflatoxins. *J. Assoc. Off. Anal. Chem.*, **65**, 1520-1523

3. PRINCIPLE

Sterigmatocystin is completely degraded by an excess of sodium hypochlorite solution followed by addition, after dilution, of acetone to destroy potential hazardous dichloro derivatives.

4. HAZARDS

4.1 *From sterigmatocystin*

Sterigmatocystin is carcinogenic in mice and rats, and has been classified by the IARC Monographs working groups as being possibly carcinogenic to man. It is recommended that handling of sterigmatocystin should be carried out in a ventilated enclosure. Gloves and other protective clothing must be worn for all operations involving handling of these compounds or their solutions.

Moreover, it has been demontrated that solutions of aflatoxins in chloroform can diffuse through latex and vinyl gloves. Should gloves come into contact with a mycotoxin solution, they should be changed as quickly as possible, to reduce the risk of contact of the mycotoxin with the skin.

To avoid problems of dispersion due to electrostatic effects, sterigmatocystin in the powder form should be handled using cotton gloves.

4.2 *Other hazards*

Sodium hypochlorite is a strong oxidizing agent; care must be taken not to mix it with concentrated reducing agents or other oxidizable substances. In the case of skin contact with corrosive chemicals, wash the skin under flowing water for at least 15 min.

The haloform reaction occurs when acetone is added to hypochlorite solutions. It is necessary to reduce the hypochlorite strength to less than 5°Cl before the addition of acetone to avoid an exothermic reaction.

5. REAGENTS

5.1 *For destruction*

Methanol	Technical grade
Acetone	Technical grace
Dichloromethane	Technical grade
Sodium hypochlorite solution (5°Cl, see Notes 1 to 3)	Technical grade

NOTE 1: *Hypochlorite solution*

It must be remembered that solutions of sodium hypochlorite tend to deteriorate. It is therefore essential to check their available chlorine content before use.

Note that strength of sodium hypochlorite solutions may be given as weight/weight or weight/volume. This is an additional reason for estimating the concentration of available chlorine.

NOTE 2: *Definition*

Percent (%) available chlorine = mass of chlorine in grams, liberated by acidifying 100 g of sodium hypochlorite solution. The available chlorine may also be expressed as °Cl, which corresponds to the volume of chlorine, in litres, liberated by one litre (liquid) or one kilogram (solid) of commercial bleach when treated with hydrochloric acid, e.g., a 1 mol/L solution of hypochlorite corresponds to 22.4°Cl.

NOTE 3: *Determination of hypochlorite strength*

The sodium hypochlorite solution used for this determination should contain not less than 25 g and not more than 30 g of available chlorine per litre.
Assay: Pipette 10.00 mL of sodium hypochlorite solution into a 100 mL volumetric flash and fill to the mark with distilled water. Pipette 10.00 mL of the resulting solution into a conical flask which already contains 50 mL of distilled water, 1 g of potassium iodide and 12.5 mL of acetic acid (2 mol/L), rinse and titrate with 0.05 mol/L sodium thiosulfate, using starch as indicator. 1 mL sodium thiosulfate (0.05 mol/L) corresponds to 3.545 mg available chlorine.

5.2 *For analysis*

Glacial acetic acid	Analytical grade
Water	Deionized
Methanol	Analytical grade

6. APPARATUS

Usual laboratory equipment and the following items:
Rotary evaporator
HPLC system equipped with a UV spectrophotometric detector
Ultrasound bath

7. PROCEDURE

Two hundred micrograms of sterigmatocystin in 4 mL of methanol are completely degraded by treatment with 5 mL of a 5°Cl hypochlorite solution for 1 h. Further treatment with acetone is effected to remove potential mutagenic compounds.

7.1 *Solid compound*

7.1.1 Estimate the amount of sterigmatocystin and thus, the volume of methanol to be added (10 mL of methanol per 0.5 mg of sterigmatocystin).

7.1.2 Add the required amount of methanol and sonicate until complete dissolution of the sterigmatocystin. Heating up to 50°C may be required.

7.1.3 For each 8 mL of solution, add 10 mL of 5°Cl sodium hypochlorite solution.

7.1.4 Allow to react for 1 h.

7.1.5 Add acetone to achieve 5% concentration.

7.1.6 Allow to react for 5 min.

7.1.7 Test for efficiency of degradation (see Section 8).

7.1.8 Dilute and discard.

7.2 *Aqueous solutions*

7.2.1 To each volume of aqueous solution, add two volumes of sodium hypochlorite solution.

7.2.2 Proceed as in 7.1.4 to 7.1.8.

7.3 *Solutions in volatile organic solvents*

7.3.1 Estimate the total amount of sterigmatocystin to be degraded.

7.3.2 Remove solvent using a rotary evaporator under reduced pressure.

7.3.3 Proceed as in 7.1.2 to 7.1.8.

7.4. *Solutions in DMSO or DMF*

7.4.1 For each volume of DMF or DMSO, add 2 volumes of water.

7.4.2 Extract three successive time with equal volumes of dichloromethane. Pool dichloromethane extracts, then dry over anhydrous sodium sulfate. Filter off sodium sulfate and rinse it with 1 volume of dichloromethane.

7.4.3 Evaporate to dryness using a rotary evaporator under reduced pressure, making sure that no solvent remains (increasing the temperature or the vacuum may be required).

7.4.4 Proceed as in 7.1.2 to 7.1.8.

7.5 *Glassware and protective clothing*

7.5.1 Rinse glassware with a small portion of methanol to dissolve possible residues of sterigmatocystin and treat solution as in 7.3.

7.5.2 Immerse glassware or protective clothing in a 5°Cl hypochlorite solution and leave in contact for 1 h.

7.5.3 Add an amount of acetone equivalent to 5% of total volume. Allow to react for at least 5 min.

7.5.4 Dilute decontaminating solution and discard.

7.6 *Spills*

7.6.1 Isolate the area and put on protective clothing.

7.6.2 Collect spills of liquid on a tissue or solid on a tissue wetted with methanol and treat them as for protective clothing.

7.6.3 Cover the area with a 5°Cl sodium hypochlorite solution; allow to react for 1 h.

7.6.4 Absorb on tissue and add to 5% acetone in water. Allow to react for at least 5 min.

7.6.5 If desired, check the surface for completeness of removal by wiping it with absorbent material moistened with methanol and analysing the wipe.

7.7 *Thin-layer chromatography plates*

Spray the plate with sodium hypochlorite solution (at least 5°Cl) and allow to stand for 1 h. Then spray with a 5% aqueous solution of acetone and allow to react for 5 min. The efficiency of degradation can be checked by scraping the plate and eluting with a suitable solvent, which can be injected directly in HPLC.

8. ANALYSIS FOR COMPLETENESS OF DEGRADATION

8.1 Take an aliquot from 7.1.8, 7.2.2, 7.3.3, 7.4.4, 7.5.4 or 7.6.5 and acidify to pH 5 – 6 using glacial acetic acid.

8.2 Bubble nitrogen through the solution for at least 1 min.

8.3 Analyse by HPLC UV using, for example, the following conditions:
Column: 25 cm × 3.6 mm i.d. Partisil ODS-2 10 μm
Precolumn: ODS 5 cm × 3.6 mm i.d.
Eluent: Methanol:water (80:20)
Flow rate: 1.5 mL/min
Detector: UV 243 nm

9. SCHEMATIC REPRESENTATION OF THE PROCEDURE

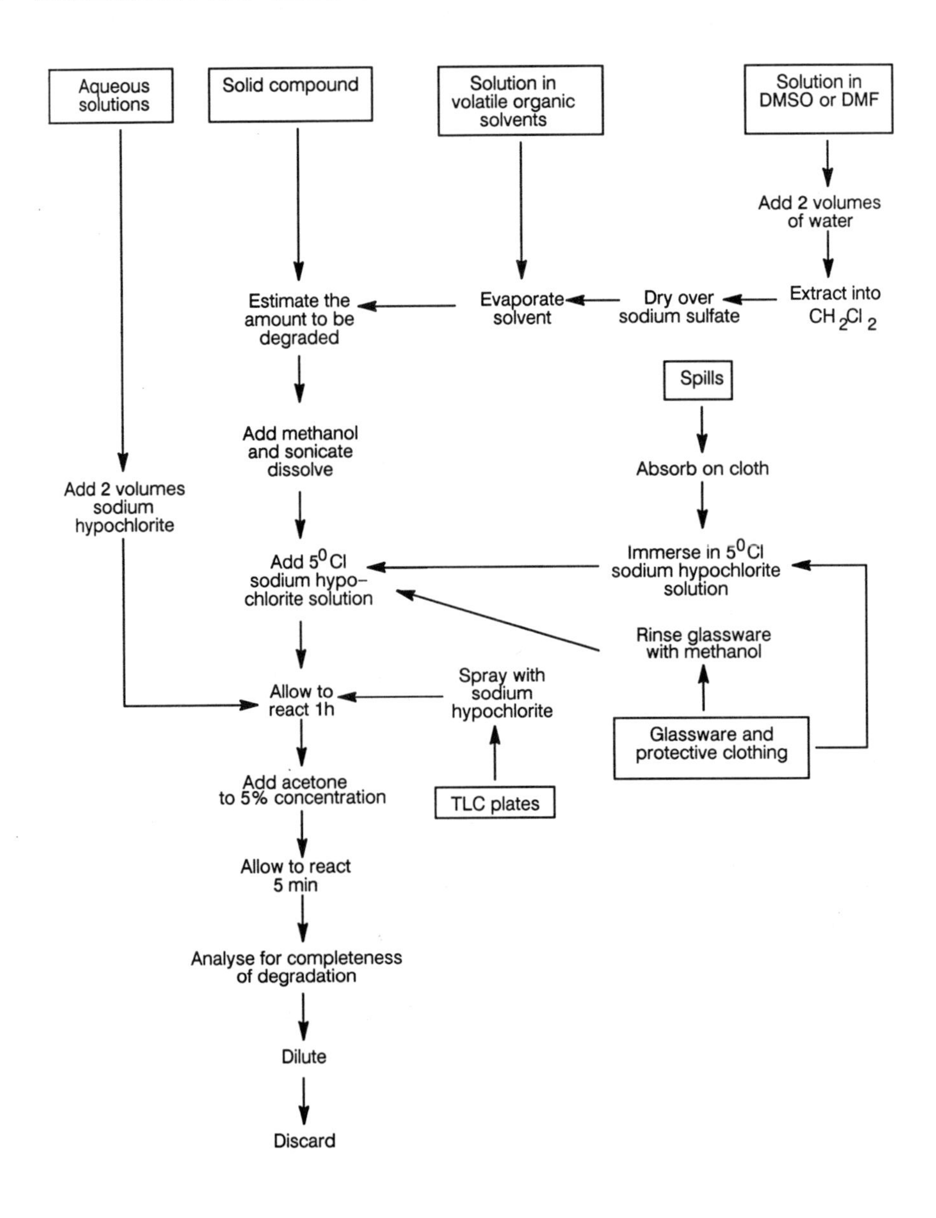

10. ORIGIN OF METHOD

M. Castegnaro and J. Michelon
IARC
150, cours Albert-Thomas
69372 Lyon Cedex 08, France

M. De Méo and M. Laget
Laboratoire de Microbiologie
Faculté de pharmacie
27, bd Jean-Moulin
13385 Marseille Cedex 05
France

Contact point: M. Castegnaro

Method 4

Destruction of patulin in animal litter using ammoniation

1. SCOPE AND FIELD OF APPLICATION

The method specifies a procedure for the treatment of animal litter.

The method has been collaboratively tested for litter contaminated with 500 μg patulin. It affords better than 99.5% destruction.

The residues of degradation by this method were tested for mutagenic activity on *Salmonella typhimurium* strains TA1530, TA1535, TA98 and TA100 with and without metabolic activation. No mutagenic activity was detected.

2. REFERENCE

Ellis, J.R., McCalla, T.M. & Norstadt, F.A. (1980) Soil effects on patulin disappearance and the effect of ammonia on patulin phytotoxicity. *Soil Science*, **129**, 371-375

3. PRINCIPLE

Litter is treated with ammonia and autoclaved without pre-evacuation.

4. HAZARDS

4.1 *From patulin*

Patulin is immunosuppressive and there is limited evidence of its carcinogenicity in experimental animals. However, mutagenicity is positive in the Ames test. It is recommended that handling of patulin should always be carried out in a ventilated enclosure. Gloves and other protective clothing must be worn for all operations involving handling of patulin or its solutions.

Moreover, it has been demonstrated that solutions of aflatoxins in chloroform can diffuse through latex and vinyl gloves. Should gloves come into contact with a mycotoxin solution, they should therefore be changed as quickly as possible, to reduce the risk of contact of the mycotoxin with the skin.

To avoid problems of dispersion due to electrostatic effects, patulin in the powder form should be handled using cotton gloves.

4.2 *From ammonia*

Ammonia solution is corrosive and gives off toxic ammonia gas.

5. REAGENTS

5.1 *For degradation*

Ammonia solution	5% (w/w)

5.2 *For analysis*

Dichloromethane	Analytical grade
Sodium sulfate	Anhydrous
Methanol	HPLC grade
Water	Deionized

6. APPARATUS

Usual laboratory equipment and the following items.

Suitable HPLC system equipped with a UV spectrophotometric detector.

Autoclave: Type normally used for the sterilization of equipment in biological laboratories.

7. PROCEDURE

7.1 Spread the contaminated litter on a suitable tray to a maximum depth of about 5 cm.

7.2 Sprinkle the litter with 5% (w/w) ammonia solution (16 mL/10 g litter).

7.3 Autoclave the tray and its contents for 20 minutes at 128 – 130°C.

NOTE: Do not pre-evacuate the autoclave. This would remove ammonia and lead to incomplete decontamination.

7.4 Cool to room temperature.

7.5 Analyse the litter for completeness of decontamination, using the method outlined in Section 8.

8. ANALYSIS FOR COMPLETENESS OF DEGRADATION

8.1 Place 10 g of autoclaved litter from 7.4 in a flask and add 50 mL of dichloromethane.

8.2 Stir this mixture for 30 min, filter, and separate the layers.

8.3 Dry the lower, organic layer over anhydrous sodium sulfate.
Transfer the dried solution to a suitable flask.

8.4 Remove the solvent by evaporation in a rotary evaporator under reduced pressure.

8.5 Take up the residue in about 2 mL of methanol.

8.6 Analyse by HPLC using, for example, the following conditions:
Column: 250 cm × 4.6 mm i.d. Microsorb C8
Mobile phase: methanol:water (10:90)
Flow rate: 1 ml/min
If a variable wavelength detector is available, set it at 276 nm, otherwise a fixed wavelength detector operating at 254 nm should be satisfactory.

9. ORIGIN OF METHOD

G. Lunn and E. B. Sansone
NCI-Frederick Cancer Research and Development Center
P.O. Box B
Frederick, MD 21702, USA
Contact point: E. B. Sansone

Method 5

Destruction of patulin and some other mycotoxins using potassium permanganate under alkaline conditions

1. SCOPE AND FIELD OF APPLICATION

This method specifies a procedure for the treatment of the following laboratory wastes: solid compounds (6.1); solutions in volatile organic solvents (6.2); glassware (6.3); aqueous solutions (6.4) and spills (6.5).

The method has been collaboratively tested using 100 μg of patulin. In this case it affords better than 99.9% decontamination.

It has also been tested by the originators of the method for the degradation of 60 μg of sterigmatocystin in 1 mL acetonitrile, 400 μg citrinin 0.2 mL acetonitrile, 400 μg OA in 0.5 mL acetonitrile and 60 μg aflatoxin B_1, B_2, G_1 or G_2 in 0.5 mL acetonitrile. It was found to achieve >99.9% degradation fo all the toxins. The efficiency was further confirmed for OA by two other laboratories.

The residues obtained by this method for degradation of OA, citrinin, AFB_1, AFB_2, AFG_1, AFG_2 and sterigmatocystin have been tested for mutagenic activity using *Salmonella typhimurium* strains TA97, TA98, TA100 and TA102 with and without metabolic activation. No mutagenic activity was detected.

2. PRINCIPLE

Degradation is effected by oxidation with potassium permanganate (0.3 mol/L) in sodium hydroxide (2 mol/L) solution.

3. HAZARDS

3.1 *From mycotoxins*

Aflatoxins are carcinogenic to humans. Citrinin is nephrotoxic and there is limited evidence of its carcinogenicity in experimental animals; OA is teratogenic and carcinogenic to experimental animals; sterigmatocystin is carcinogenic in mice and rats and has been classified as possibly carcinogenic to humans; patulin is immunosuppressive and there is limited evidence of its carcinogenicity in experimental animals, but mutagenicity is positive in the Ames test. It is recommended that handling of mycotoxins should always be carried out in a ventilated enclosure. Gloves and other protective clothing must be worn for all operations involving handling of these compounds or their solutions.

Moreover, it has been demonstrated that solutions of aflatoxins in chloroform can diffuse through latex and vinyl gloves. Should gloves come into contact with a mycotoxin solution, they should therefore be changed as quickly as possible, to reduce the risk of contact of the mycotoxin with the skin.

To avoid problems of dispersion due to electrostatic effects, mycotoxins in the powder form should be handled using cotton gloves.

3.2 *Other hazards*

Potassium permanganate is a strong oxidizing agent, and care must be taken not to mix it with concentrated reducing agents. Sodium hydroxide is corrosive. In the case of skin contact with a corrosive agent, wash the skin under flowing water for at least 15 min.

4. REAGENTS

4.1 *For degradation*

Potassium permanganate	Technical grade
Potassium permanganate solution	0.3 mol/L solution in 2 mol/L sodium hydroxide prepared daily
Sodium hydroxide	Technical grade
Sodium hydroxide solution	2 mol/L, aqueous
Dichloromethane	Technical grade

4.2 *For analysis*

Acetonitrile	HPLC grade
Methanol	HPLC grade
Water	Deionized
Sodium disulfite $Na_2S_2O_5$ (CAS R.N. 7681-57-4)	Analytical grade
Sodium disulfite solution	2 mol/L, aqueous
Sodium sulfate	Analytical grade, anhydrous
Hydrochloric acid	Analytical grade, 12 mol/L
Dichloromethane	Analytical grade
Phosphoric acid	Analytical grade
Phosphoric acid	0.25 mol/L, aqueous

5. APPARATUS

Usual chemical laboratory equipment and the following items:

HPLC system equipped with a spectrofluorimetric and a UV spectrophotometric detector
Rotary evaporator

6. PROCEDURE

Ten mL of 0.3 mol/L potassium permanganate in 2 mol/L sodium hydroxide will degrade 400 µg patulin, 300 µg of sterigmatocystin or AFB_1 or AFB_2 or AFG_1 or AFG_2, 2 mg citrinin or OA, in 3 hours. However, it should be noted that other components in waste may react with potassium permanganate turning the purple/green colour to brown.

6.1 *Solid compounds*

6.1.1 Dissolve the mycotoxin (400 µg of patulin, 300 µg of sterigmatocystin, AFB_1, AFB_2, AFG_1 or AFG_2, 2 mg of citrinin or OA) in 5 mL acetonitrile.

6.1.2 To each 5 mL of solution add 10 mL of 0.3 mol/L $KMnO_4$ in 2 mol/L NaOH.

6.1.3 Allow to react for at least 3 hours with stirring and make sure that the purple or green colour remains. If not, add more potassium permanganate.

6.1.4 For each 10 mL of $KMnO_4$/NaOH solution added, add 10 mL of 2 mol/L sodium disulfite solution.

6.1.5 Analyse for completeness of degradation as described in Section 7.

6.1.6 Discard.

6.2 *Solutions in volatile organic solvents*

6.2.1 Remove solvent(s) by evaporation to dryness, using a rotary evaporation under reduced pressure.

6.2.2 Proceed as in 6.1.1 to 6.1.6.

6.3 *Glassware*

6.3.1 Rinse glassware with small portions of dichloromethane (five times).

6.3.2 Combine the rinses and evaporate dichloromethane on a rotary evaporator under reduced pressure.

6.3.3 Proceed as in 6.1.1 to 6.1.6.

6.4 *Aqueous solutions*

6.4.1 Dilute, if necessary, with water, to $\leqslant$ 200 µg per mL water and add solid NaOH to bring to 2 mol/L.

6.4.2 Add, with stirring, solid potassium permanganate to 0.3 mol/L.

6.4.3 Proceed as in 6.1.3 to 6.1.6.

6.5 *Spills*

6.5.1 Isolate the area and put on suitable protective clothing.

6.5.2 Absorb a liquid spill on a tissue or solid on a tissue wetted with dichloromethane and immerse in a bath containing 0.3 mol/L $KMnO_4$ in 2 mol/L NaOH.

6.5.3 Cover the area with an excess of a 0.3 mol/L $KMnO_4$ in 2 mol/L NaOH solution and allow to react for 3 hours.

6.5.4 Absorb the solution on a tissue and immerse in a 2 mol/L solution of sodium disulfite.

NOTE: Check the pH of the solution. If it is acidic, make it alkaline with sodium hydroxide.

6.5.5 Rinse the area with a 2 mol/L solution of sodium disulfite.

6.5.6 If desired, check the surface for completeness of removal by wiping it with an absorbent material moistened with methanol.

7. ANALYSIS FOR COMPLETENESS OF DEGRADATION

7.1 Take an aliquot from 6.1.4, 6.2.2, 6.3.3 and 6.4.3.

7.2 Acidify to about pH 2 – 3 using 12 mol/L HCl.

7.3 Extract three times with an equal volume of dichloromethane and pool dichloromethane extracts.

7.4 Dry extract over anhydrous sodium sulfate and filter off sodium sulfate.

7.5 Evaporate to near dryness using a rotary evaporator under reduced pressure, then to complete dryness under a gentle stream of nitrogen gas.

7.6 Take up residue in 500 μL of HPLC solvent.

7.7 Analyse by HPLC using, for example, the following conditions:

7.7.1 Patulin:
Column: 25 cm × 3.6 mm reverse phase Nucleosil 5 μm ODS
Eluent: water:acetonitrile (90:10)
Flow rate: 1.0 mL/min
Detector: UV spectrophotometer 275 nm

7.7.2 Sterigmatocystin
Column: 25 cm × 3.6 mm i.d. 5 μm Nucleosil C_{18}
Eluent: methanol:water (80:20)
Flow rate: 1.0 mL/min
Detector: UV spectrophotometer 243 nm

7.7.3 Aflatoxins
Column: 25 cm × 3.6 mm i.d. 10 μm ODS 2
Eluent: water:methanol:acetonitrile (2:1:1)
Flow rate: 1.0 mL/min
Detector: spectrofluorimeter excitation 360 nm, emission 440 nm

7.7.4 OA and citrinin
Column: 25 cm × 3.6 mm i.d. 5 μm Nucleosil C_{18}
Eluent: acetonitrile: 0.25 mol/L aqueous H_3PO_4 (75:25)
Flow rate: 1.0 mL/min
Detector: Spectrofluorimetry
excitation 330 nm, emission 480 nm for citrinin
excitation 340 nm, emission 465 nm for OA
UV spectroscopy 330 nm for citrinin

8. SCHEMATIC REPRESENTATION OF THE PROCEDURE

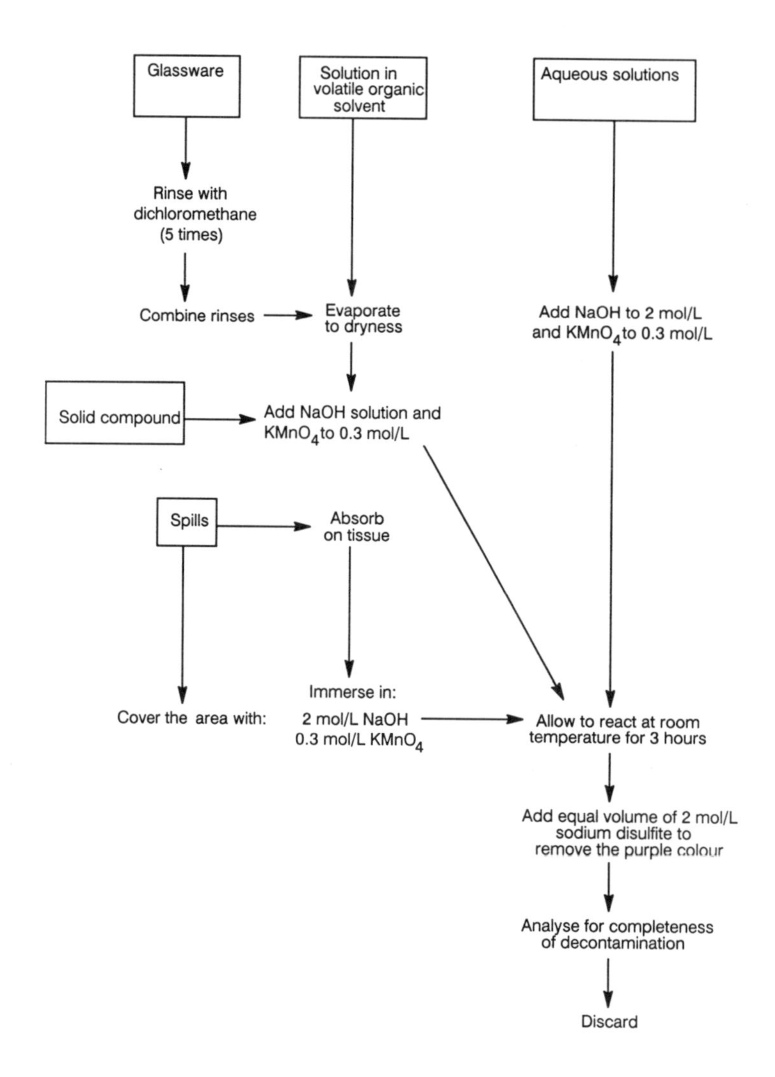

9. ORIGIN OF METHOD

M. Castegnaro
IARC
150, cours Albert-Thomas
69372 Lyon Cedex 08
France

J.-M. Fremy and
E. Gleizes
Section of Natural Toxins
CNEVA-LCHA
43, rue de Dantzig
75015 Paris
France

M. De Méo and M. Laget
Laboratoire de
Microbiologie
Faculté de pharmacie
27, bd Jean-Moulin
13385 Marseille Cedex 05
France

Contact point: M. Castegnaro

Method 6

Degradation of patulin using ammoniation

1. SCOPE AND FIELD OF APPLICATION

This method specifies a procedure for the destruction of patulin in the following wastes: solid compounds (7.1); aqueous solutions (7.2); solutions in volatile organic solvents (7.3) and glassware (7.4).

The method has been collaboratively tested using 400 μg patulin in 200 μL acetonitrile. It affords better than 99.8% degradation.

The residues obtained by this method for degradation of patulin have been tested for mutagenic activity using *Salmonella typhimurium* strains TA97, TA98, TA100 and TA102 with and without metabolic activation. No mutagenic activity was detected.

2. REFERENCE

Ellis, J.R., McCalla, T.M. & Norstadt, F.A. (1980) Soil effects on patulin disappearance and the effect of ammonia on patulin phytotoxicity. *Soil Science*, **129**, 371-375

3. PRINCIPLE

Decontamination is effected by treatment with ammonia in an autoclave.

4. HAZARDS

4.1 *From patulin*

Patulin is immunosuppressive and there is limited evidence of its carcinogenicity in experimental animals. However, mutagenicity is positive in the Ames test. It is recommended that handling of patulin should always be carried out in a ventilated enclosure. Gloves and other protective clothing must be worn for all operations involving handling of this compound or its solutions.

Moreover, it has been demonstrated that solutions of aflatoxins in chloroform can diffuse through latex and vinyl gloves. Should gloves come into contact with a mycotoxin solution, they should therefore be changed as quickly as possible, to reduce the risk of contact of the mycotoxin with the skin.

To avoid problems of dispersion due to electrostatic effect, patulin in the powder form should be handled using cotton gloves.

4.2 *From ammonia*

Ammonia solution is corrosive and gives off toxic ammonia gas.

5. REAGENTS

5.1. *For decontamination*

Ammonia solution	5% and 10% (w/w)

5.2 *For analysis*

Water	Deionized
Hydrochloric acid	Analytical grade, 12 mol/L
Hydrochloric acid	1 mol/L, aqueous
Sodium sulfate	Analytical grade, anhydrous
Ethyl acetate	Analytical grade
Acetonitrile	HPLC grade

6. APPARATUS

Usual laboratory equipment and the following items:

Autoclave
HPLC system equipped with an UV spectrophotometric detector
Rotary evaporator
Ultrasound bath

7. PROCEDURE

1000 μg of patulin are degraded by autoclaving at 120°C with 100 mL 5% ammonia for 15 min.

7.1 *Solid compound*

7.1.1 For each 100 μg of patulin, add at least 10 mL of 5% ammonia solution. *Do not close container tightly.*

7.1.2 Sonicate for about 1 min to improve solubilization.

7.1.3 Place in an autoclave and heat under pressure to reach 120°C.

7.1.4 Allow to react for 15 min.

7.1.5 Allow to come to room temperature.

7.1.6 Test for efficiency of degradation (see Section 8).

7.1.7 Dilute and discard.

7.2 *Aqueous solutions*

7.2.1 Adjust to <200 μg/10 mL, then add an equal volume of 10% ammonia solution.

7.2.2 Proceed as in 7.1.2 to 7.1.7.

7.3 *Solutions in volatile organic solvents*

7.3.1 Evaporate solvent using a rotary evaporator under reduced pressure.

7.3.2 Proceed as in 7.1.1 to 7.1.7.

7.4 *Glassware*

7.4.1 Rinse five times with ethyl acetate.

7.4.2 Treat rinse as in 7.3.

8. ANALYSIS FOR COMPLETENESS OF DEGRADATION

8.1 Take an aliquot from 7.1.7, 7.2.2, 7.3.2 or 7.4.2 and acidify to about pH 5–6 using the HCl solution.

8.2 Extract three times with an equal volume of ethyl acetate and pool ethyl acetate extracts.

8.3 Dry extract over anhydrous sodium sulfate and filter off sodium sulfate.

8.4 Evaporate to near dryness using a rotary evaporator under reduced pressure, then to complete dryness under a gentle stream of nitrogen.

8.5 Take up residue in 500 μL of water:acetonitrile mixture (90:10) (v/v).

8.6 Analyse by HPLC with UV detection by using, for example, the following conditions:
Column: 25 cm × 3.6 mm i.d. Nucleosil 5 μm ODS-2
Eluent: water:acetonitrile (90:10)
Flow rate: 1.0 mL/min
Detector: UV $\lambda = 275$ nm

9. SCHEMATIC REPRESENTATION OF THE PROCEDURE

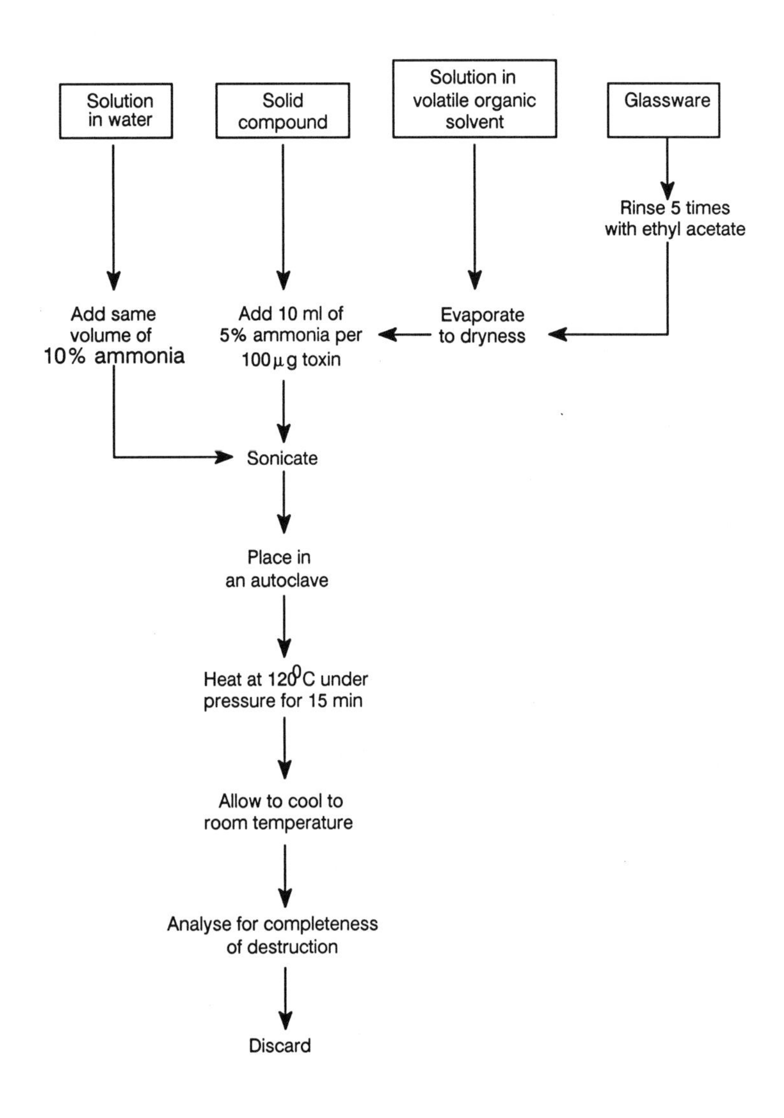

10. ORIGIN OF METHOD

J.-M. Fremy and E. Gleizes
Section of Natural Toxins
CNEVA-LCHA
43, rue de Dantzig
75015 Paris

Contact point: J.-M. Fremy

M. De Méo and M. Laget
Laboratoire de Microbiologie
Faculté de Pharmacie
27, bd Jean-Moulin
13385 Marseille Cedex 05
France

Appendix A

Nomenclature and chemical and physical data for the mycotoxins considered

1. Citrinin

Nomenclature

Chemical Abstracts Services Registry Number: 518-75-2

Chemical Abstracts Name: (3R-*trans*)-4,6-Dihydro-8-hydroxy-3,4,5-trimethyl-6-oxo-6-3*H*-2-benzopyran-7-carboxylic acid

IUPAC Name: (3*R*,4*S*)-4,6-Dihydro-8-hydroxy-3,4,5-trimethyl-6-oxo-6-3*H*-2-benzopyran-7-carboxylic acid

Other name: Antimycin, S-52

Molecular and structural information

Molecular formula: $C_{13}H_{14}O_5$

Molecular weight: 250.25

Structural formula:

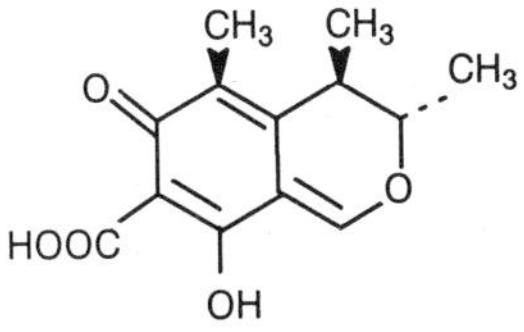

Physical properties and spectral data

Description: Lemon yellow glistening crystals (Ambrose & De Eds, 1946). Yellow odourless, crystalline solid (Pohland *et al.*, 1982). Lemon yellow needles from ethanol (Windholz, 1983) or methanol (Haese, 1963). Yellow needles from methanol (Weast, 1985)

Melting point: 168 – 170°C (Haese, 1963). Melts with decomposition at 170 – 172°C (Hetherington & Raistrick, 1931)
170 – 173°C after drying for 1 hour at 60°C (Pohland *et al.*, 1982). Decomposes at 175°C (Cartwright *et al.*, 1949; Neely *et al.*, 1972; Windholz, 1983).

Specific rotation: $[\alpha]_D^{11}$ – 37° (Weast, 1985) $[\alpha]_D^{18}$ – 37.4° (Cartwright *et al.*, 1949; Windholz, 1983)

Solubility: Practically insoluble in water. Soluble in alcohol, dioxane and dilute alkali (Windholz, 1983). Solubility at 25°C in petroleum ether (35° – 60°) 0.56 mg/L; diethyl ether, 5.78 mg/L; ethanol, 7.08 mg/L; ethyl acetate, 21.1 mg/L; acetone, 49.4 mg/L; benzene, 28.58 mg/L; chloroform, 77.4 mg/L (Ambrose & De Eds, 1946).

Spectral data: Ultraviolet (Cram, 1950; Neely *et al.*, 1972; Pohland *et al.*, 1982), fluorescence (Neely *et al.*, 1972), phosphorescence (Neely *et al.*, 1972), infrared (Kovàč *et al.*, 1961; Pouchert, 1981; Pohland *et al.*, 1982), proton magnetic resonance (Mathieson & Whalley, 1964; Barber & Staunton, 1980; Pohland *et al.*, 1982), C-13 nuclear magnetic resonance (Barber & Staunton, 1980) and mass spectral data (Pohland *et al.*, 1982) have been investigated.

2. Ochratoxin A

Nomenclature

Chemical Abstracts Services Registry Number: 303-47-9

Chemical Abstracts name: L-Phenylalanine, *N*-[(5-chloro-3,4-dihydro-8-hydroxy-3-methyl-1-oxo-1*H*-2-benzopyran-7-yl)carbonyl]-,(*R*)-

IUPAC name: *N*-[[(3*R*)-5-chloro-8-hydroxy-3-methyl-1-oxo-7-isochromanyl] carbonyl]-3-phenyl-L-alanine

Other names: (–)-*N*-[(5-chloro-8-hydroxy-3-methyl-1-oxo-7-isochromanyl)carbonyl]-3-phenylalanine

Molecular and structural information

Molecular formula: $C_{20}H_{18}ClNO_6$

Molecular weight: 403.8

Structural formula:

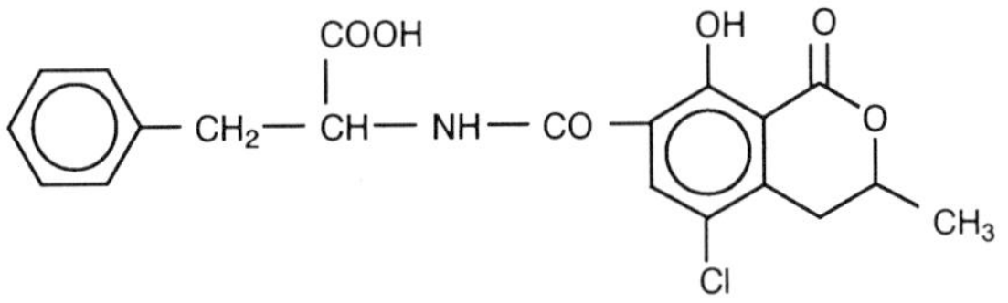

COOH
CH$_2$— CH— NH— CO
OH O
O
CH$_3$
Cl

Physical properties and spectral data

Description: Crystals from xylene (Windholz, 1983); white odourless crystalline solid (Pohland *et al.*, 1982).

Melting point: 159°C (recrystallized from benzene – hexane) (Natori *et al.*, 1970)
169°C (recrystallized from xylene) (Van der Merwe *et al.*, 1965 a,b)
168 – 173°C, after drying for 1 hour at 60°C (Pohland *et al.*, 1982).

Specific rotation: $[\alpha]_D^{20}$ – 118° (*c* 1.1 mmol/L in chloroform) (Van der Merwe *et al.*, 1965 a,b)
$[\alpha]_D^{21}$ – 46.8° (*c* 2.65 mmol/L in chloroform) (Pohland *et al.*, 1982).

Circular dichroism: $\Delta\varepsilon_{(\lambda 395)}$ 0, $\Delta\varepsilon_{(\lambda 327)}$ – 1.45, $\Delta\varepsilon_{(\lambda 300)}$ 0, $\Delta\varepsilon_{(\lambda 285)}$ + 0.2, $\Delta\varepsilon_{(\lambda 275)}$ 0, $\Delta\varepsilon_{(\lambda 235)}$ – 6.37, $\Delta\varepsilon_{(\lambda 218)}$ 0
$c = 0.2613$ mmol/L in methanol; $\theta = 22$°C; cell length 2 mm (Pohland *et al.*, 1982).

Solubility: Moderately soluble in organic solvents (chloroform, ethanol and methanol).

Spectral data: Ultraviolet (Van der Merwe *et al.*, 1965 a,b; Steyn & Holzapfel, 1967; Pohland *et al.*, 1982), infrared (Van der Merwe *et al.*, 1965 a,b; Steyn & Holzapfel, 1967; Pohland *et al.*, 1982).

Nuclear magnetic resonance (Pohland *et al.*, 1982) and mass (NIH-EPA Information System, 1982; Pohland *et al.*, 1982) spectra have been investigated.

3. Patulin

Nomenclature

Chemical Abstracts Services Registry Number: 149-29-1

Chemical Abstracts and IUPAC name: 4-hydroxy-4*H*-furo[3,2-*c*]pyran-2(6*H*)-one

Other names: anhydro-3-hydroxymethylene-tetrahydro-γ-pyrene-2-carboxylic acid; clairformin; clavacin; clavatin; claviformin; (2,4-dihydroxy-2*H*-pyran-3-(6*H*)ylidene) acetic acid, 3,4-lactone; expansin; expansine; mycoin; mycoin C; mycoin C3; penicidin; terinin; gigantin; leucopin; mycoine C3; mycosin; penatin

Molecular and structural information

Molecular formula: $C_7H_6O_4$

Molecular weight: 154.1

Structural formula:

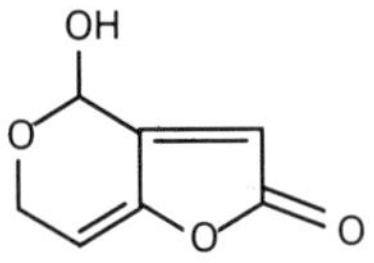

OH
O
O O

Physical properties and spectral data

Description: White odourless crystalline solid (Pohland *et al.*, 1982). Prisms or plates from ether or chloroform (Windholz, 1983; Weast, 1985)

Melting point: 111°C (Birkinshaw *et al.*, 1943). 109.5 – 110.5°C (Bergel *et al.*, 1944). 109°C (Brack, 1947). 105 – 108°C, after drying for 1 hour at 60°C (Pohland *et al.*, 1982), 111°C (Windholz, 1983; Weast, 1985)

Specific rotation: $[\alpha]_D^{21}$ −6.2° (c=6.489 mmol/L in chloroform) (Pohland *et al.*, 1982)

Solubility: Soluble in water and the common organic solvents except petroleum ether (Birkinshaw *et al.*, 1943). Very soluble in ethyl or amyl acetate (Windholz, 1983). Soluble in water, alcohol, ether, acetone, benzene (Weast, 1985) and also chloroform (Waksman *et al.*, 1942)

Spectral data: Ultraviolet (Bergel *et al.*, 1944: Pohland *et al.*, 1982), infrared (Pohland *et al.*, 1982), nuclear magnetic resonance (Pohland *et al.*, 1982) and mass (Pohland *et al.*, 1982; ICIS Chemical Information, 1984) spectra have been investigated.

4. Sterigmatocystin

Nomenclature

Chemical Abstracts Services Registry Number: 10048-13-2

Chemical Abstracts name: 3*a*,12*c*-Dihydro-8-hydroxy-6-methoxy-7*H*-furo-[3',2':4,5]furo[2,3-*c*]xanthen-7-one

Molecular and structural information

Molecular formula: $C_{18}H_{12}O_6$

Molecular weight: 324.3

Structural formula:

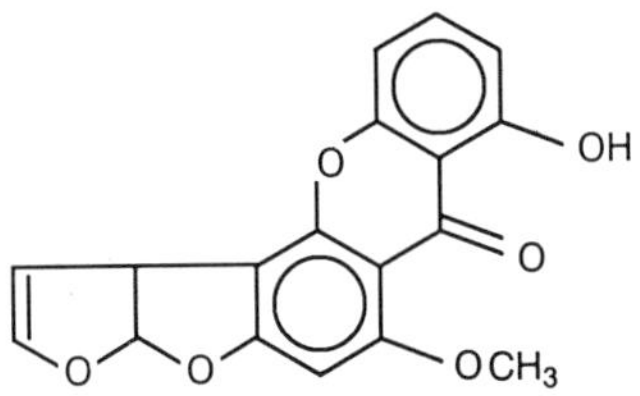

OH
O
O
O
O
OCH_3

Physical properties and spectral data

From Pohland *et al.* (1982) (I) and Quintavalla (1986) (II) unless otherwise specified

Description:	Yellow odourless crystalline solid
Melting point:	242 – 244°C, after drying for 1 hour at 60°C (I). 246°C (II)
Specific rotation:	$[\alpha]_D^{21}$ – 369.0° (c = 3.084 mmol/L in chloroform) (I). $[\alpha]_D^{20}$ – 387° (in chloroform) (II)
Circular dichroism:	$\Delta\varepsilon_{(\lambda 375)}$ 0, $\Delta\varepsilon_{(\lambda 325)}$ – 2.58, $\Delta\varepsilon_{(\lambda 297)}$ 0, $\Delta\varepsilon_{(\lambda 290)}$ + 1.42, $\Delta\varepsilon_{(\lambda 280)}$ 0.90, $\Delta\varepsilon_{(\lambda 270)}$ + 1.42, $\Delta\varepsilon_{(\lambda 263)}$ 0, c = 2.923 mmol/L in methanol; θ = 22°C; cell length 2 mm
Solubility:	Insoluble in water and strong alkali; sparingly soluble in most organic solvents; readily soluble in chloroform, pyridine and dimethyl sulfoxide (IARC, 1976)
Spectral data:	Ultraviolet in methanol and ethanol, infrared, nuclear magnetic resonance and mass spectra have been investigated.

Appendix B

Biological degradation techniques and chemical reactions of the mycotoxins considered

Reviews of the methods which have been assayed to degrade various mycotoxins have been published by Müller (1982, 1983)

Compound	Product	Reaction	Reference
Citrinin	"Phenol A"	Reflux with a 10% sodium hydroxide solution for 4 h under nitrogen	Barber *et al.* (1981)
Citrinin		Oxidation by 5 mol/L chromic acid in presence of 3 mol/L sulfuric acid for 1.5 h at 100°C	Barber *et al.* (1981)
Citrinin		Reaction with pyridine-acetic anhydride. Allow to stand 15 min at room temperature.	Hald & Krogh (1973)
Citrinin		Reflux in hexane (68°C) or 95% ethanol (78°C) solution initiates decomposition	Neely *et al.* (1972)
Citrinin		Autoclaving of maize grain at 121°C for 20 min reduces content by 80%.	Jackson & Ciegler (1978)
Citrinin		70% breakdown by treatment with phosphate buffer pH 11 at 38°C for 24 h	Müller (1983)
Citrinin		Treatment with phosphate buffer pH 2 at 21°C for 24 h reduces level to 60%.	Müller (1983)
Citrinin	Ochratoxin A + other	Transformation by intracellular factors released by *P. viridicatum*	Patterson & Damoglou (1987)
Citrinin	Dehydro-citrinin	Hydrogenation in methanol in presence of palladium-charcoal catalyst.	Brown *et al.* (1949) Afzal & Ashat (1981)

Compound	Product	Reaction	Reference
Citrinin	Methylcitrinin	Treatment with methyl sulfate in a saturated sodium hydrogen carbonate solution.	Brown *et al.* (1949)
Citrinin	CO_2 + 2 products	Heating in presence of 2N H_2SO_4 leads to complete decomposition	Hetherington & Raistrick (1931)
Patulin	Not specified but with loss of biological activity	Treatment with alkali	Windholz (1983)
Patulin		Heating for 20 min at 80°C results only in 25-50% loss of patulin. In model system, at pH 2.3 to 6.9, glutathione reduces the level of patulin.	Scott & Somers (1968)
Patulin		Acetylation by a 7:3 mixture of acetic anhydride:pyridine	Meyer (1982)
Patulin		Destabilization by compounds containing sulfhydryl groups.	Valetrisco *et al.* (1983)
Patulin		Patulin is degraded in presence of bisulfite.	Adam & Koller (1979)
Patulin		The heat stability of patulin depends on the pH. It is greater at pH 3.4 than at pH 5.5.	Lovett & Peeler (1973)
Patulin		100% patulin disappears at pH 7 or 8, at 25°C in 190 h	Brackett & Marth (1979a)
Patulin		Toxicity of patulin in wheat germ is eliminated by ammonia treatment.	Ellis *et al.*, (1980)
Patulin		Vitamin C reduces patulin content in buffer solutions pH 3.5.	Brackett & Marth (1979b)
Patulin		Patulin is very stable in buffer at pH 3.4 to 6.5.	Lovett & Peeler (1973) Adam & Koller (1979) Brackett & Marth (1979a,b)
Patulin		Patulin reacts with thiols (cysteine, glutathione).	Ciegler *et al.* (1976) Lieu & Bullerman (1978)

Compound	Product	Reaction	Reference
Patulin		More extensive degradation takes place when heated with NaOH than when treated with cold NaOH. Patulin is hydrolysed when heated with 2N H_2SO_4 for 6 h. Catalytic hydrogenation in ethanolic solution can occur in presence of palladium-norite catalyst.	Birkinshaw *et al.* (1943)
Patulin	Oxime Methyl ester Benzoate	Reaction with hydroxylamine hydrochloride in 0.2 M HCl Reflux in anhydrous ether: methyl iodide (10:4) in presence of silver oxide. Treatment with benzoic acid in pyridine. Catalytic hydrogenation in 95% ethanol-water or acetic acid in presence of palladium-charcoal.	Bergel *et al.* (1944)
Patulin		High temperature (60 – 90°C) short time pasteurization reduces patulin levels but does not achieve complete destruction of the toxin.	Wheeler *et al.* (1987)
Sterigmatocystin	Bright yellow fluorescent compound	Treatment with $AlCl_3$ (20%) in ethanol and heating at 120°C for 20 min.	Purchase & Pretorius (1973)
Sterigmatocystin	Orange red	Treatment with trifluorofluorescent acetic acid compound	Holzapfel *et al.* (1966)
Sterigmatocystin	Aflatoxin B_1	An extract prepared from *Aspergillus flavus* ATCC 5517/A228 contains a protein capable of converting sterigmatocystin into AFB_1. The activity of the enzyme isolated is maximum at 28°C and pH 8 and is enhanced by Zn^{2+}, Co^{2+} and Mn^{2+}.	Mashaly *et al.* (1988)
Sterigmatocystin		Irradiation with a 450 W medium-pressure mercury lamp at 0°C for 20 min in presence of oxygen and hydrolysis with 0.1 N HCl 100°C for 1 h.	Büchi *et al.* (1982)
Sterigmatocystin		Reduction with sodium borohydride.	Chen *et al.* (1977)
Ochratoxin A	L-Phenylalanine and an acid of formula $C_{11}H_9O_5Cl$	Acid hydrolysis in 6 N HCl for 30 h under reflux.	Van der Meerwe *et al.* (1965a, b)
Ochratoxin A	*O*-Methylochratoxin A methyl ester	Treatment with an excess ethereal solution of diazomethane for 16 h.	Van der Meerwe *et al.* (1965a, b)

Compound	Product	Reaction	Reference
Ochratoxin A		Autoclaving of two raw white bean samples for 1 h at 121°C reduced their OA content by 13.3 and 9.5%.	Harwig *et al.* (1974)
Ochratoxin A		Cooking the mash in beer brewing reduces the OA content by less than 10%.	Chu *et al.* (1975)
Ochratoxin A		Autoclaving oatmeal and rice cereal for 30 min at 121°C reduces the OA content by 83 and 89%. An increased water content reduces the degradation efficiency.	Trenk *et al.* (1971)
Ochratoxin A		Treatment at 198 – 210°C reduces the OA content in coffee beans and wheat by averages of 83% and 89% respectively in periods ranging from 5 to 20 min.	Levi *et al.* (1974)
Ochratoxin A		During the manufacture of bread, no OA drop could be detected after cooking for 25 min at 220°C, but a 62% reduction was achieved in biscuits cooked for 5.5 min at 180°C. The higher the water content, the lower the loss.	Osborne (1979)
Ochratoxin A		Roasting slaughter products at 150 – 160°C for 12 min reduces the OA content by an average of 20%.	Josefsson & Möller (1980)
Ochratoxin A, citrinin and sterigmatocystin		Treatment with 2% NH_3 solution reduces the chemically identifiable content of OA and citrinin but sterigmatocystin seems unaffected.	Chelkowski *et al.* (1981)
Ochratoxin A		Treatment of maize for 96 h at 70°C with 5% NH_3 in air reduces its OA content by 95% but the OA content of the kidney of animals fed with this treated maize was only slightly affected	Madsen *et al.* (1980, 1983)
Ochratoxin		Ammoniation long enough to completely break down OA suppresses its toxicity in chick embryo tests.	Chelkowski *et al.* (1982)
Ochratoxin A	Ochratoxin A methyl or ethyl ester	Reaction with 14% BF_3 in methanol for 5 min at boiling. Reaction with 16% BF_3 in anhydrous ethanol in an ice bath.	Nesheim (1969)
Ochratoxin A		Four metabolites are formed by resting cell medium of *Phenylobacterium immobile*	Wegst & Lingens (1983)
Ochratoxin A		80 to 90% of OA in coffee is removed by roasting.	Gallaz & Stalder (1976)

References

Adam, R. & Koller, W.D. (1979) Untersuchungen über den Einfluss von Hydrogensulfitionen auf das Schimmelpilzgift Patulin. 1. Versuch in wässrigen Lösungen. *Dtsch. Lebensm. Rdsch.*, **75**, 254-256.

Afzal, S.M. & Ashraf, M. (1981) Synthesis of 3,4,5-trimethyl-6,8-dihydroxy-7-isopentyl-3,4-dihydroisocoumarin. *Pakistan J. Sci.*, **33**, 80-83.

Ambrose, A.M. & De Eds, F. (1946) Some toxicological and pharmacological properties of citrinin. *J. Pharmacol. Exp. Ther.*, **88**, 173-186.

Barber, J. & Staunton, J. (1980) New insights into polyketide metabolism; the use of protium as a tracer in the biosynthesis of citrinin by penicillium citrinum. *J. Chem. Soc., Perkin Trans.* I, **9**, 2244-2248.

Barber, J., Carter, R.H., Garson, M.J. & Staunton, J. (1981) The biosynthesis of citrinin by penicillium citrinum. *J. Chem. Soc., Perkin Trans.* I, **9**, 2577-2583.

Bergel, F., Morrison, A.L., Moss, A.R. & Rinderknecht, H. (1944) An antibacterial substance from *Aspergillus clavatus. J. Chem. Soc.*, 415-421.

Bicking, M.K. & Kniseley, R.N. (1980) Size exclusion chromatographic analysis of refuse-derived fuel for mycotoxins. *Analyt. Chem.*, **52**, 2164-2168.

Birkinshaw, J.H., Michael, S.E., Bracken, A. & Raistrick, H. (1943) Patulin in the common cold. II. Biochemistry and chemistry. *Lancet*, **245**, 625-630.

Boorman, G. (1988) *NTP Technical Report on the Toxicology and Carcinogenesis Studies of Ochratoxin A.* NTP TR358, NIH publication No. 88-2813.

Brack, A. (1947) Isolierung von Gentisinalkohol neben Patulin aus dem Kulturfiltrat eines Penicillium-Stammes und über einige Derivate des Gentisinalkohols. *Helv. Chim. Acta*, **30**, 1-8.

Brackett, R.E. & Marth, E.H. (1979a) Stability of patulin at pH 6.0-8.0 and 25°C. *Z. Lebensm. Unters. Forsch.*, **169**, 92-94.

Brackett, R.E. & Marth E.H. (1979b) Ascorbic acid and ascorbate cause disappearance of patulin from buffer solutions and apple juice. *J. Food Prot.*, **42**, 864-866.

Brown, J.P., Robertson, A., Whalley, W.B. & Cartwright, N.J. (1949) The chemistry of fungi. Part V. The constitution of citrinin. *J. Chem. Soc.*, 867-879.

Büchi, G., Fowler, K.W. & Nadzan, A.M. (1982) Photochemical oxidation of aflatoxin B_1 and sterigmatocystin: synthesis of guanine-containing adducts. *J. Am. Chem. Soc.*, **104**, 544-547.

Candlish, A.A.G., Stimson, W.H. & Smith, J.E. (1988) Determination of ochratoxin A by monoclonal antibody-based enzyme immunoassay. *J. Ass. Off. Anal. Chem.*, **71**, 961-964.

Cartwright, N.J., Robertson, A. & Whalley, W.B. (1949) The chemistry of fungi. Part VII. Synthesis of citrinin and dehydrocitrinin, *J. Chem. Soc.*, 1563-1567.

Castegnaro, M., Hunt, D.C., Sansone, E.B., Schuller, P.L., Siriwardana, M.G., Telling, G.M., Van Egmond, H.P. & Walker, E.A., eds (1980) *Laboratory Decontamination and Destruction of Aflatoxins B_1, B_2, G_1, G_2 in Laboratory Wastes* (IARC Scientific Publications No. 37), Lyon, International Agency for Research on Cancer.

Castegnaro, M., Chernozemsky, I.N., Hietanen, E. & Bartsch, H. (1990) Are mycotoxins risk factors for endemic nephropathy and associated urothelial cancers? *Arch. Geschwulstforsch.*, **60**, 295-303.

Chalam, R.V. & Stahr, H.M. (1979) Thin-layer chromatographic determination of citrinin. *J. Ass. Off. Anal. Chem.*, **62**, 570-572.

Chelkowski, J., Golinski, P., Godlewska, B., Radomyska, W., Szebiotko, K. & Wiewiorowska, M. (1981) Mycotoxins in cereal grain. Part IV. Inactivation of ochratoxin A and other mycotoxins during ammoniation. *Nahrung*, **25**, 631-637.

Chelkowski, J., Szebiotko, K., Golinski, P., Buchowski, M., Godlewska, B., Radomyska, W. & Wiewioroska, M. (1982) Mycotoxins in cereal grain. Part 5. Changes of cereal grain biological value after ammoniation and mycotoxins (ochratoxins) inactivation. *Nahrung*, **26**, 631-637.

Chen, P.N., Kingston, D.G.I. & Vercellotti, J.R. (1977) Reduction of sterigmatocystin and versicolorin A hemiacetals with sodium borohydride. *J. Org. Chem.*, **42**, 3599-3605.

Chu, F.S., Chang, C.C., Ashoor, S.H. & Prentice, N. (1975) Stability of aflatoxin B_1 and ochratoxin A in brewing. *Appl. Microbiol.*, **29**, 313-316.

Chu, F.S., Chang, F.C.C. & Hinsdill, R.D. (1976) Production of antibody against ochratoxin A. *Appl. Environ. Microbiol.*, **31**, 831-835.

Ciegler, A., Beckwith, A.C. & Jackson, L.K. (1976) Teratogenicity of patulin and patulin adducts formed with cysteine. *Appl. Microbiol.*, **31**, 664-667.

Cohen, H. & Lapointe, M. (1986) Determination of ochratoxin A in animal feed and cereal grains by gas-liquid chromatography with fluorescence detection. *J. Ass. Off. Anal. Chem.*, **69**, 957-959.

Cram, D.J. (1950) Mold metabolites. V. The stereochemistry and ultraviolet absorption spectrum of citrinin. *J. Chem. Soc.*, **72**, 1001-1002.

Ehman, J. & Gaucher, G.M. (1977) Quantitation of patulin pathway metabolites using gas-liquid chromatography. *J. Chromatog.*, **132**, 17-26.

Ellis, J.R., McCalla, T.M. & Norstadt, F.A. (1980) Soil effects on patulin disappearance and the effect of ammonia on patulin phytotoxicity. *Soil Science*, **129**, 371-375.

Forbito, P.R. & Babsky, N.E. (1985) Rapid liquid chromatographic determination of patulin in apple juice. *J. Ass. Off. Anal. Chem.*, **68**, 950-951.

Frisvad, J.C. (1987) High performance liquid chromatographic determination of profiles of mycotoxins and other secondary metabolites. *J. Chromatog.*, **392**, 333-347.

Frohlich, A.A., Marquardt, R.R. & Bernatsky, A. (1988) Quantitation of ochratoxin A: Use of reverse phase thin-layer chromatography for sample clean-up followed by liquid chromatography or direct fluorescence measurement. *J. Ass. Off. Anal. Chem.*, **71**, 949-953.

Gallaz, L. & Stalder, R. (1976) Ochratoxin A im Kaffee. *Chem. Mikrobiol. Technol. Lebensm.*, **4**, 147-149.

Gimeno, A. (1984) Determination of citrinin in corn and barley on thin layer chromatographic plates impregnated with glycolic acid. *J. Ass. Off. Anal. Chem.*, **67**, 194-196.

Gimeno, A. & Martins, M.L. (1983) Rapid thin layer chromatographic determination of patulin, citrinin, and aflatoxin in apples and pears, and their juices and jams. *J. Ass. Off. Anal. Chem.*, **66**, 85-91.

Haese, G. (1963) Über Antimycin. *Arch. Pharmacol.*, **296**, 227-232.

Hald, B. & Krogh, P. (1973) Analysis and chemical confirmation of citrinin in barley. *J. Ass. Off. Anal. Chem.*, **56**, 1440-1443.

Harwig, J., Chen, Y.K. & Collings-Thompson, D.L. (1974) Stability of ochratoxin A in beans during canning. *J. Can. Inst. Food Sci. Technol.*, **7**, 288-289.

Hetherington, A.C. & Raistrick, H. (1931) Studies in biochemistry of micro-organisms. Part XIV. On the production and chemical constitution of a new yellow colouring matter, citrinin, produced from glucose by *Penicillium citrinium Thom. Trans. Roy. Soc., London*, B220, 269-295.

Holzapfel, C.W., Purchase, I.F.H., Steyn, P.S. & Gouws, L. (1966) The toxicity and chemical assay of sterigmatocystin, a carcinogenic mycotoxin, and its isolation from two new fungal sources. *South African Med. J.*, **40**, 1100-1101.

Hunt, D.C., Bourdon, A.T. & Crosby, N.T. (1978) Use of high performance liquid chromatography for the identification and estimation of zearalenone, patulin and penicillic acid in food. *J. Sci. Fd. Agric.*, **29**, 239-244.

Hurst, W.J., Snyder, K.P. & Martin, R.A., Jr (1987) High performance liquid chromatographic determination of the mycotoxins patulin, penicillic acid, zearalenone and sterigmatocystin in artificially contaminated cocoa beans. *J. Chromatog.*, **392**, 389-396.

ICIS Chemical Information System (1984) Carbon-13 NMR spectral search system (CNMR), mass spectral search system (MSSS), infrared spectral search system (IRSS), information system for hazardous organics in water (ISHOW) and environmental face (ENVIROFATE), Washington DC, Information Consultants Inc.

International Agency for Research on Cancer (1976) *IARC Monographs on the Evaluation of Carcinogenic Risk of Chemicals to Man.* Volume 10, *Some Naturally Occurring Substances*, Lyon.

International Agency for Research on Cancer (1983) *IARC Monographs on the Evaluation of Carcinogenic Risk of Chemicals to Humans.* Volume 31, *Some Food Additives, Feed Additives and Naturally Occurring Substances*, Lyon.

International Agency for Research on Cancer (1986) *IARC Monographs on the Evaluation of Carcinogenic Risk of Chemicals to Humans.* Volume 40, *Some Naturally Occurring and Synthetic Food Components, Furocoumarins and Ultraviolet Radiations*, Lyon.

International Agency for Research on Cancer (1987) *IARC Monographs on the Evaluation of Carcinogenic Risk of Chemicals to Humans*, Supplement 7, *Overall Evaluations of Carcinogenicity: An Updating of IARC Monographs Volumes 1 to 42*, Lyon.

Jackson, L.K. & Ciegler, A. (1978) Production and analysis of citrinin in corn. *Appl. Environ. Microbiol.*, **36**, 408-411.

Josefsson, B.G.E. & Möller, T.E. (1980) Heat stability of ochratoxin A in pig products. *J. Sci. Fd. Agric.*, **31**, 1313-1315.

Kawamura, O., Sato, S., Kajii, H., Nagayama, S., Ohtani, K., Chiba, J. & Ueno, Y. (1989) A sensitive enzyme-linked immunosorbent assay of ochratoxin A based on monoclonal antibodies. *Toxicon*, **27**, 887-897.

Kovàč, S., Nemec, P. & Betina, V. (1961) Chemical structure of citrinin. *Nature*, **190**, 1104-1105.

Lepom, P. (1986a) Determination of sterigmatocystin in feed by high performance liquid chromatography with column switching. *J. Chromatog.*, **354**, 518-523.

Lepom, P. (1986b) Simultaneous determination of the mycotoxins citrinin and ochratoxin A in wheat and barley by high performance liquid chromatography. *J. Chromatog.*, **355**, 335-339.

Levi, C.P., Trenk, H.L. & Mohr, H.K. (1974) Study of the occurrence of ochratoxin A in green coffee beans. *J. Ass. Off. Anal. Chem.*, **57**, 866-870.

Lieu, F.Y. & Bullerman, L.B. (1978) Binding of patulin and penicillic acid to glutathione and cysteine and toxicity of the resulting adducts. *Milchwissenschaft*, **33**, 16-20.

Lovett, J. & Peeler, J.T. (1973) Effect of pH on the thermal destruction kinetics of patulin in aqueous solution. *J. Food Sci.*, **38**, 1094-1095.

Madsen, A., Elling, F., Hald, B., Mortensen, H.P. & Winther, P. (1980) The effect on pigs of ochratoxin A-contaminated barley treated with ammonia. In: Nielsen, N.C., Høgh, P. & Bille, N., eds, *Proceedings of IPVS Congress, Copenhagen*, p. 287.

Madsen, A., Hald, B. & Mortensen, H.P. (1983) Feeding experiments with ochratoxin A contaminated barley for bacon pigs. 3. Detoxification by ammoniation heating + NaOH, or autoclaving. *Acta Agric. Scand.*, **33**, 171-175.

Marti, L.R., Wilson, D.M. & Evans, B.D. (1978) Determination of citrinin in corn and barley. *J. Ass. Off. Anal. Chem.*, **61**, 1353-1358.

Mashaly, R.I., Habid, S.L., El-Deeb, S.A., Salem, M.H. & Safwat, M.M. (1988) Isolation, purification and characterisation of enzyme(s) responsible for conversion of sterigmatocystin to aflatoxin B_1. *Z. Lebensm. Unters. Forsch.*, **186**, 118-124.

Mathieson, D.W. & Whalley, W.B. (1964) The chemistry of fungi. Part XLIV. The conformation of citrinin. *J. Chem. Soc.*, 4640-4641.

Meyer, R.A. (1982) Zur Bestimmung von Patulin in Lebensmitteln. Vorschlag einer Standard Methode. *Die Nahrung*, **26**, 337-342.

Morgan, M.R.A. (1989) Mycotoxin immunoassays (with special reference to ELISAs). *Tetrahedron*, **45**, 2237-2249.

Morgan, M.R.A., McNerney, R. & Chan, H.W.S. (1983) Enzyme-linked immunosorbent assay of ochratoxin A in barley. *J. Ass. Off. Anal. Chem.*, **66**, 1481-1484.

Morgan, M.R.A., McNerney, R., Chan, H.W.S. & Anderson, P.H. (1986) Ochratoxin A in pig kidney determined by enzyme-linked immunosorbent assay (ELISA). *J. Sci. Food Agric.*, **37**, 475-480.

Müller, H.M. (1982) Entgiftung von Mykotoxinen. I. Physikalische Verfahren. *Übersicht. Tierernahr.*, **10**, 95-122.

Müller, H.M. (1983) Entgiftung von Mykotoxinen. II. Chemische Verfahren und Reaktion mit Inhaltstoffen von Futtermitteln. *Übersicht. Tierernahr.*, **11**, 47-80.

Natori, S., Sakaki, S., Kurata, H., Udagawa, S., Ichinoe, M., Saito, M. & Umeda, M. (1970) Chemical and cytotoxicity survey on the production of ochratoxins and penicillic acid by *Aspergillus ochraceus Wihl. Chem. Pharm. Bull.*, **18**, 2259-2268.

Neely, W.C., Ellis, S.P., Davis, N.D. & Diener, U.L. (1972) Spectroanalytical parameters of fungal metabolites. I. Citrinin. *J. Ass. Off. Anal. Chem.*, **55**, 1122-1127.

Nesheim, S. (1969) Isolation and purification of ochratoxin A and B and preparation of their methyl and ethyl esters. *J. Ass. Off. Anal. Chem.*, **52**, 975-979.

NIH/EPA Chemical Information System (1982) *Mass Spectral Search System*, Washington DC, CIS Project, Information Sciences Corporation.

Ogawa, H., Suzuki, J., Ochiai, S., Nishijima, M., Naoi, Y. & Nishima, T. (1984) Determination of penicillic acid and patulin in commercial fruit juice by gas-liquid chromatography. *Kenku Nenpo*, **35**, 224-230.

Osborne, B.G. (1979) Reverse-phase high performance liquid chromatography determination of ochratoxin A in flour and bakery products. *J. Sci. Fd. Agric.*, **30**, 1065-1070.

Patterson, M.F. & Damoglou, A.P. (1987) Conversion of the mycotoxin citrinin into dihydrocitrinin and ochratoxin A by *Penicillium viridicatum. Appl. Microbiol. Biotechnol.*, **27**, 574-578.

Phillips, R.D., Hayes, A.W. & Berndt, W.O. (1980) High performance liquid chromatographic analysis of the mycotoxin citrinin and its application to biological fluids. *J. Chromatog.*, **190**, 419-427.

Pohland, A.E., Schuller, P.L., Steyn, P.S. & Van Egmond, H.P. (1982) Physicochemical data for some selected mycotoxins. *Pure Appl. Chem.*, **54**, 2219-2284.

Pouchert, C.J., ed. (1981) *The Aldrich Library of Infrared Spectra*, 3rd ed., Milwaukee, WI, Aldrich Chemical Co., p. 321B.

Price, K.R. (1979) A comparison of two quantitative mass spectrometry methods for the analysis of patulin in apple juice. *Biomed. Mass Spectrom*, **6**, 573-574.

Purchase, I.F.H. & Pretorius, M.E. (1973) Sterigmatocystin in coffee beans. *J. Ass. Off. Anal. Chem.*, **56**, 225-226.

Quintavalla, S. (1986) Le sterigmatocistine. *Industria Conserve*, **61**, 34-41.

Rousseu, D.M., Candlish, A.A.G., Slegers, G.A., van Peteghem, C.H., Stinson, W.H. & Smith, J.E. (1987) Detection of ochratoxin A in porcine kidneys by monoclonal antibody-based radioimmunoassay. *Appl. Environ. Microbiol.*, **53**, 514-518.

Salhab, A.S., Russell, G.F., Coughlin, J.R. & Hsieh, D.P.H. (1976) Gas-liquid chromatography and mass spectrometric ion selective detection of sterigmatocystin in grains. *J. Ass. Off. Anal. Chem.*, **59**, 1037-1044.

Schmidt, R., Mondani, J., Ziegenhagen, E. & Dose, K. (1981) High performance liquid chromatography of the mycotoxin sterigmatocystin and its application to the analysis of moldy rice for sterigmatocystin. *J. Chromatog.*, **207**, 435-438.

Scott, P.M. & Somers, E. (1968) Stability of patulin and penicillic acid in fruit juices and flour. *J. Agric. Fd. Chem.*, **16**, 483-485.

Skierska, D. & Martinek, W. (1984) Determination of patulin in apple juice by gas-liquid chromatography with an electron capture detector. *Rocz. Panstw. Zakl. Hig.*, **35**, 69-75.

Steyn, P.S. & Holzapfel, C.W. (1967) The synthesis of ochratoxins A and B metabolites of *Aspergillus ochraceus Wihl. Tetrahedron*, **23**, 4449-4461.

Stoloff, L., Castegnaro, M., Scott, P., O'Neill, I.K. & Bartsch, H., eds (1982) *Environmental Carcinogens, Selected Methods of Analysis.* Volume 5, *Some Mycotoxins* (IARC Scientific Publications No. 44), Lyon, International Agency for Research on Cancer.

Suzuki, T., Hoshino, Y., Kikuchi, Y., Nose, N. & Watanabe, A. (1976) Determination of sterigmatocystin by gas-liquid chromatography with an electron capture detector. *Shokuhin Eiseigaku Zasshi*, **17**, 253-257.

Tranthan, A.L. & Wilson, D.M. (1984) Fluorometric screening method for citrinin in corn, barley and peanuts. *J. Ass. Off. Anal. Chem.*, **67**, 37-38.

Trenk, H.L., Butz, M.E. & Chu, F.S. (1971) Production of ochratoxins in different cereal products by *Aspergillus ochraceus. Appl. Microbiol.*, **21**, 1032-1035.

Valetrisco, M., Ruggieri, G. & Niola, I. (1983) Studio sperimentale su alcuni fattori possibili destabilizzanti della patuline. *Industrie Alimentari*, 953-956.

Van der Merwe, K.J., Steyn, P.S., Fourie, L., Scott, de B. & Theron, J.J. (1965a) Ochratoxin A, a toxic metabolite produced by *Aspergillus ochraceus Wihl. Nature*, **205**, 1112-1113.

Van der Merwe, K.J., Steyn, P.S. & Fourie, L. (1965b) Mycotoxins. Part II. The constitution of ochratoxins A, B and C, metabolites of *Aspergillus ochraceus Wihl. J. Chem. Soc.*, 7083-7088.

Waksman, S.A., Horning, E.S. & Spencer E.L. (1942) The production of two antibacterial substances, fumigacin and clavacin. *Science*, **96**, 202-203.

Weast, R.C., ed. (1985) *CRC Handbook of Chemistry and Physics*, 66th ed., Boca Raton, FL, CRC Press, p. C-213 (citrinin).

Wegst, W. & Lingens, F. (1983) Bacterial degradation of ochratoxin A. *FEMS Microbiol. Lett.*, **17**, 341-344.

Wheeler, J.L., Harrison, M.A. & Koehler, P.E. (1987) Presence and stability of patulin in pasteurized apple cider. *J. Food Sci.*, **52**, 479-480.

Williams, S., ed. (1984) *Official Methods of Analysis of the Association of Official Analytical Chemists*, fourteenth edition, Arlington, VA, Association of Official Analytical Chemists.

Windholz, M., ed. (1983) *The Merck Index*, 10th ed., Rahway, NJ, Merck, p. 331.

PUBLICATIONS OF THE INTERNATIONAL AGENCY FOR RESEARCH ON CANCER

Scientific Publications Series

(Available from Oxford University Press through local bookshops)

No. 1 **Liver Cancer**
1971; 176 pages (*out of print*)

No. 2 **Oncogenesis and Herpesviruses**
Edited by P.M. Biggs, G. de-Thé and L.N. Payne
1972; 515 pages (*out of print*)

No. 3 ***N*-Nitroso Compounds: Analysis and Formation**
Edited by P. Bogovski, R. Preussman and E.A. Walker
1972; 140 pages (*out of print*)

No. 4 **Transplacental Carcinogenesis**
Edited by L. Tomatis and U. Mohr
1973; 181 pages (*out of print*)

No. 5/6 **Pathology of Tumours in Laboratory Animals, Volume 1, Tumours of the Rat**
Edited by V.S. Turusov
1973/1976; 533 pages; £50.00

No. 7 **Host Environment Interactions in the Etiology of Cancer in Man**
Edited by R. Doll and I. Vodopija
1973; 464 pages; £32.50

No. 8 **Biological Effects of Asbestos**
Edited by P. Bogovski, J.C. Gilson, V. Timbrell and J.C. Wagner
1973; 346 pages (*out of print*)

No. 9 ***N*-Nitroso Compounds in the Environment**
Edited by P. Bogovski and E.A. Walker
1974; 243 pages; £21.00

No. 10 **Chemical Carcinogenesis Essays**
Edited by R. Montesano and L. Tomatis
1974; 230 pages (*out of print*)

No. 11 **Oncogenesis and Herpesviruses II**
Edited by G. de-Thé, M.A. Epstein and H. zur Hausen
1975; Part I: 511 pages
Part II: 403 pages; £65.00

No. 12 **Screening Tests in Chemical Carcinogenesis**
Edited by R. Montesano, H. Bartsch and L. Tomatis
1976; 666 pages; £45.00

No. 13 **Environmental Pollution and Carcinogenic Risks**
Edited by C. Rosenfeld and W. Davis
1975; 441 pages (*out of print*)

No. 14 **Environmental *N*-Nitroso Compounds. Analysis and Formation**
Edited by E.A. Walker, P. Bogovski and L. Griciute
1976; 512 pages; £37.50

No. 15 **Cancer Incidence in Five Continents, Volume III**
Edited by J.A.H. Waterhouse, C. Muir, P. Correa and J. Powell
1976; 584 pages; (*out of print*)

No. 16 **Air Pollution and Cancer in Man**
Edited by U. Mohr, D. Schmähl and L. Tomatis
1977; 328 pages (*out of print*)

No. 17 **Directory of On-going Research in Cancer Epidemiology 1977**
Edited by C.S. Muir and G. Wagner
1977; 599 pages (*out of print*)

No. 18 **Environmental Carcinogens. Selected Methods of Analysis. Volume 1: Analysis of Volatile Nitrosamines in Food**
Editor-in-Chief: H. Egan
1978; 212 pages (*out of print*)

No. 19 **Environmental Aspects of *N*-Nitroso Compounds**
Edited by E.A. Walker, M. Castegnaro, L. Griciute and R.E. Lyle
1978; 561 pages (*out of print*)

No. 20 **Nasopharyngeal Carcinoma: Etiology and Control**
Edited by G. de-Thé and Y. Ito
1978; 606 pages (*out of print*)

No. 21 **Cancer Registration and its Techniques**
Edited by R. MacLennan, C. Muir, R. Steinitz and A. Winkler
1978; 235 pages; £35.00

No. 22 **Environmental Carcinogens. Selected Methods of Analysis. Volume 2: Methods for the Measurement of Vinyl Chloride in Poly(vinyl chloride), Air, Water and Foodstuffs**
Editor-in-Chief: H. Egan
1978; 142 pages (*out of print*)

No. 23 **Pathology of Tumours in Laboratory Animals. Volume II: Tumours of the Mouse**
Editor-in-Chief: V.S. Turusov
1979; 669 pages (*out of print*)

No. 24 **Oncogenesis and Herpesviruses III**
Edited by G. de-Thé, W. Henle and F. Rapp
1978; Part I: 580 pages, Part II: 512 pages (*out of print*)

Prices, valid for September 1991, are subject to change without notice

No. 25 **Carcinogenic Risk. Strategies for Intervention**
Edited by W. Davis and C. Rosenfeld
1979; 280 pages (*out of print*)

No. 26 **Directory of On-going Research in Cancer Epidemiology 1978**
Edited by C.S. Muir and G. Wagner
1978; 550 pages (*out of print*)

No. 27 **Molecular and Cellular Aspects of Carcinogen Screening Tests**
Edited by R. Montesano, H. Bartsch and L. Tomatis
1980; 372 pages; £29.00

No. 28 **Directory of On-going Research in Cancer Epidemiology 1979**
Edited by C.S. Muir and G. Wagner
1979; 672 pages (*out of print*)

No. 29 **Environmental Carcinogens. Selected Methods of Analysis. Volume 3: Analysis of Polycyclic Aromatic Hydrocarbons in Environmental Samples**
Editor-in-Chief: H. Egan
1979; 240 pages (*out of print*)

No. 30 **Biological Effects of Mineral Fibres**
Editor-in-Chief: J.C. Wagner
1980; **Volume 1:** 494 pages; **Volume 2:** 513 pages; £65.00

No. 31 ***N*-Nitroso Compounds: Analysis, Formation and Occurrence**
Edited by E.A. Walker, L. Griciute, M. Castegnaro and M. Börzsönyi
1980; 835 pages (*out of print*)

No. 32 **Statistical Methods in Cancer Research. Volume 1. The Analysis of Case-control Studies**
By N.E. Breslow and N.E. Day
1980; 338 pages; £20.00

No. 33 **Handling Chemical Carcinogens in the Laboratory**
Edited by R. Montesano *et al.*
1979; 32 pages (*out of print*)

No. 34 **Pathology of Tumours in Laboratory Animals. Volume III. Tumours of the Hamster**
Editor-in-Chief: V.S. Turusov
1982; 461 pages; £39.00

No. 35 **Directory of On-going Research in Cancer Epidemiology 1980**
Edited by C.S. Muir and G. Wagner
1980; 660 pages (*out of print*)

No. 36 **Cancer Mortality by Occupation and Social Class 1851-1971**
Edited by W.P.D. Logan
1982; 253 pages; £22.50

No. 37 **Laboratory Decontamination and Destruction of Aflatoxins B_1, B_2, G_1, G_2 in Laboratory Wastes**
Edited by M. Castegnaro *et al.*
1980; 56 pages; £6.50

No. 38 **Directory of On-going Research in Cancer Epidemiology 1981**
Edited by C.S. Muir and G. Wagner
1981; 696 pages (*out of print*)

No. 39 **Host Factors in Human Carcinogenesis**
Edited by H. Bartsch and B. Armstrong
1982; 583 pages; £46.00

No. 40 **Environmental Carcinogens. Selected Methods of Analysis. Volume 4: Some Aromatic Amines and Azo Dyes in the General and Industrial Environment**
Edited by L. Fishbein, M. Castegnaro, I.K. O'Neill and H. Bartsch
1981; 347 pages; £29.00

No. 41 ***N*-Nitroso Compounds: Occurrence and Biological Effects**
Edited by H. Bartsch, I.K. O'Neill, M. Castegnaro and M. Okada
1982; 755 pages; £48.00

No. 42 **Cancer Incidence in Five Continents, Volume IV**
Edited by J. Waterhouse, C. Muir, K. Shanmugaratnam and J. Powell
1982; 811 pages (*out of print*)

No. 43 **Laboratory Decontamination and Destruction of Carcinogens in Laboratory Wastes: Some *N*-Nitrosamines**
Edited by M. Castegnaro *et al.*
1982; 73 pages; £7.50

No. 44 **Environmental Carcinogens. Selected Methods of Analysis. Volume 5: Some Mycotoxins**
Edited by L. Stoloff, M. Castegnaro, P. Scott, I.K. O'Neill and H. Bartsch
1983; 455 pages; £29.00

No. 45 **Environmental Carcinogens. Selected Methods of Analysis. Volume 6: *N*-Nitroso Compounds**
Edited by R. Preussmann, I.K. O'Neill, G. Eisenbrand, B. Spiegelhalder and H. Bartsch
1983; 508 pages; £29.00

No. 46 **Directory of On-going Research in Cancer Epidemiology 1982**
Edited by C.S. Muir and G. Wagner
1982; 722 pages (*out of print*)

No. 47 **Cancer Incidence in Singapore 1968–1977**
Edited by K. Shanmugaratnam, H.P. Lee and N.E. Day
1983; 171 pages (*out of print*)

No. 48 **Cancer Incidence in the USSR (2nd Revised Edition)**
Edited by N.P. Napalkov, G.F. Tserkovny, V.M. Merabishvili, D.M. Parkin, M. Smans and C.S. Muir
1983; 75 pages; £12.00

No. 49 **Laboratory Decontamination and Destruction of Carcinogens in Laboratory Wastes: Some Polycyclic Aromatic Hydrocarbons**
Edited by M. Castegnaro *et al.*
1983; 87 pages; £9.00

No. 50 **Directory of On-going Research in Cancer Epidemiology 1983**
Edited by C.S. Muir and G. Wagner
1983; 731 pages (*out of print*)

No. 51 **Modulators of Experimental Carcinogenesis**
Edited by V. Turusov and R. Montesano
1983; 307 pages; £22.50

No. 52 **Second Cancers in Relation to Radiation Treatment for Cervical Cancer: Results of a Cancer Registry Collaboration**
Edited by N.E. Day and J.C. Boice, Jr
1984; 207 pages; £20.00

No. 53 **Nickel in the Human Environment**
Editor-in-Chief: F.W. Sunderman, Jr
1984; 529 pages; £41.00

No. 54 **Laboratory Decontamination and Destruction of Carcinogens in Laboratory Wastes: Some Hydrazines**
Edited by M. Castegnaro *et al.*
1983; 87 pages; £9.00

No. 55 **Laboratory Decontamination and Destruction of Carcinogens in Laboratory Wastes: Some *N*-Nitrosamides**
Edited by M. Castegnaro *et al.*
1984; 66 pages; £7.50

No. 56 **Models, Mechanisms and Etiology of Tumour Promotion**
Edited by M. Börzsönyi, N.E. Day, K. Lapis and H. Yamasaki
1984; 532 pages; £42.00

No. 57 ***N*-Nitroso Compounds: Occurrence, Biological Effects and Relevance to Human Cancer**
Edited by I.K. O'Neill, R.C. von Borstel, C.T. Miller, J. Long and H. Bartsch
1984; 1013 pages; £80.00

No. 58 **Age-related Factors in Carcinogenesis**
Edited by A. Likhachev, V. Anisimov and R. Montesano
1985; 288 pages; £20.00

No. 59 **Monitoring Human Exposure to Carcinogenic and Mutagenic Agents**
Edited by A. Berlin, M. Draper, K. Hemminki and H. Vainio
1984; 457 pages; £27.50

No. 60 **Burkitt's Lymphoma: A Human Cancer Model**
Edited by G. Lenoir, G. O'Conor and C.L.M. Olweny
1985; 484 pages; £29.00

No. 61 **Laboratory Decontamination and Destruction of Carcinogens in Laboratory Wastes: Some Haloethers**
Edited by M. Castegnaro *et al.*
1985; 55 pages; £7.50

No. 62 **Directory of On-going Research in Cancer Epidemiology 1984**
Edited by C.S. Muir and G. Wagner
1984; 717 pages (*out of print*)

No. 63 **Virus-associated Cancers in Africa**
Edited by A.O. Williams, G.T. O'Conor, G.B. de-Thé and C.A. Johnson
1984; 773 pages; £22.00

No. 64 **Laboratory Decontamination and Destruction of Carcinogens in Laboratory Wastes: Some Aromatic Amines and 4-Nitrobiphenyl**
Edited by M. Castegnaro *et al.*
1985; 84 pages; £6.95

No. 65 **Interpretation of Negative Epidemiological Evidence for Carcinogenicity**
Edited by N.J. Wald and R. Doll
1985; 232 pages; £20.00

No. 66 **The Role of the Registry in Cancer Control**
Edited by D.M. Parkin, G. Wagner and C.S. Muir
1985; 152 pages; £10.00

No. 67 **Transformation Assay of Established Cell Lines: Mechanisms and Application**
Edited by T. Kakunaga and H. Yamasaki
1985; 225 pages; £20.00

No. 68 **Environmental Carcinogens. Selected Methods of Analysis. Volume 7. Some Volatile Halogenated Hydrocarbons**
Edited by L. Fishbein and I.K. O'Neill
1985; 479 pages; £42.00

No. 69 **Directory of On-going Research in Cancer Epidemiology 1985**
Edited by C.S. Muir and G. Wagner
1985; 745 pages; £22.00

No. 70 **The Role of Cyclic Nucleic Acid Adducts in Carcinogenesis and Mutagenesis**
Edited by B. Singer and H. Bartsch
1986; 467 pages; £40.00

No. 71 **Environmental Carcinogens. Selected Methods of Analysis. Volume 8: Some Metals: As, Be, Cd, Cr, Ni, Pb, Se Zn**
Edited by I.K. O'Neill, P. Schuller and L. Fishbein
1986; 485 pages; £42.00

No. 72 **Atlas of Cancer in Scotland, 1975–1980. Incidence and Epidemiological Perspective**
Edited by I. Kemp, P. Boyle, M. Smans and C.S. Muir
1985; 285 pages; £35.00

No. 73 **Laboratory Decontamination and Destruction of Carcinogens in Laboratory Wastes: Some Antineoplastic Agents**
Edited by M. Castegnaro *et al.*
1985; 163 pages; £10.00

No. 74 **Tobacco: A Major International Health Hazard**
Edited by D. Zaridze and R. Peto
1986; 324 pages; £20.00

No. 75 **Cancer Occurrence in Developing Countries**
Edited by D.M. Parkin
1986; 339 pages; £20.00

No. 76 **Screening for Cancer of the Uterine Cervix**
Edited by M. Hakama, A.B. Miller and N.E. Day
1986; 315 pages; £25.00

List of IARC Publications

No. 77 **Hexachlorobenzene: Proceedings of an International Symposium**
Edited by C.R. Morris and J.R.P. Cabral
1986; 668 pages; £50.00

No. 78 **Carcinogenicity of Alkylating Cytostatic Drugs**
Edited by D. Schmähl and J.M. Kaldor
1986; 337 pages; £25.00

No. 79 **Statistical Methods in Cancer Research. Volume III: The Design and Analysis of Long-term Animal Experiments**
By J.J. Gart, D. Krewski, P.N. Lee, R.E. Tarone and J. Wahrendorf
1986; 213 pages; £20.00

No. 80 **Directory of On-going Research in Cancer Epidemiology 1986**
Edited by C.S. Muir and G. Wagner
1986; 805 pages; £22.00

No. 81 **Environmental Carcinogens: Methods of Analysis and Exposure Measurement. Volume 9: Passive Smoking**
Edited by I.K. O'Neill, K.D. Brunnemann, B. Dodet and D. Hoffmann
1987; 383 pages; £35.00

No. 82 **Statistical Methods in Cancer Research. Volume II: The Design and Analysis of Cohort Studies**
By N.E. Breslow and N.E. Day
1987; 404 pages; £30.00

No. 83 **Long-term and Short-term Assays for Carcinogens: A Critical Appraisal**
Edited by R. Montesano, H. Bartsch, H. Vainio, J. Wilbourn and H. Yamasaki
1986; 575 pages; £48.00

No. 84 **The Relevance of *N*-Nitroso Compounds to Human Cancer: Exposure and Mechanisms**
Edited by H. Bartsch, I.K. O'Neill and R. Schulte- Hermann
1987; 671 pages; £50.00

No. 85 **Environmental Carcinogens: Methods of Analysis and Exposure Measurement. Volume 10: Benzene and Alkylated Benzenes**
Edited by L. Fishbein and I.K. O'Neill
1988; 327 pages; £35.00

No. 86 **Directory of On-going Research in Cancer Epidemiology 1987**
Edited by D.M. Parkin and J. Wahrendorf
1987; 676 pages; £22.00

No. 87 **International Incidence of Childhood Cancer**
Edited by D.M. Parkin, C.A. Stiller, C.A. Bieber, G.J. Draper, B. Terracini and J.L. Young
1988; 401 pages; £35.00

No. 88 **Cancer Incidence in Five Continents Volume V**
Edited by C. Muir, J. Waterhouse, T. Mack, J. Powell and S. Whelan
1987; 1004 pages; £50.00

No. 89 **Method for Detecting DNA Damaging Agents in Humans: Applications in Cancer Epidemiology and Prevention**
Edited by H. Bartsch, K. Hemminki and I.K. O'Neill
1988; 518 pages; £45.00

No. 90 **Non-occupational Exposure to Mineral Fibres**
Edited by J. Bignon, J. Peto and R. Saracci
1989; 500 pages; £45.00

No. 91 **Trends in Cancer Incidence in Singapore 1968–1982**
Edited by H.P. Lee , N.E. Day and K. Shanmugaratnam
1988; 160 pages; £25.00

No. 92 **Cell Differentiation, Genes and Cancer**
Edited by T. Kakunaga, T. Sugimura, L. Tomatis and H. Yamasaki
1988; 204 pages; £25.00

No. 93 **Directory of On-going Research in Cancer Epidemiology 1988**
Edited by M. Coleman and J. Wahrendorf
1988; 662 pages (*out of print*)

No. 94 **Human Papillomavirus and Cervical Cancer**
Edited by N. Muñoz, F.X. Bosch and O.M. Jensen
1989; 154 pages; £19.00

No. 95 **Cancer Registration: Principles and Methods**
Edited by O.M. Jensen, D.M. Parkin, R. MacLennan, C.S. Muir and R. Skeet
1991; 288 pages; £28.00

No. 96 **Perinatal and Multigeneration Carcinogenesis**
Edited by N.P. Napalkov, J.M. Rice, L. Tomatis and H. Yamasaki
1989; 436 pages; £48.00

No. 97 **Occupational Exposure to Silica and Cancer Risk**
Edited by L. Simonato, A.C. Fletcher, R. Saracci and T. Thomas
1990; 124 pages; £19.00

No. 98 **Cancer Incidence in Jewish Migrants to Israel, 1961–1981**
Edited by R. Steinitz, D.M. Parkin, J.L. Young, C.A. Bieber and L. Katz
1989; 320 pages; £30.00

No. 99 **Pathology of Tumours in Laboratory Animals, Second Edition, Volume 1, Tumours of the Rat**
Edited by V.S. Turusov and U. Mohr
740 pages; £85.00

No. 100 **Cancer: Causes, Occurrence and Control**
Editor-in-Chief L. Tomatis
1990; 352 pages; £24.00

List of IARC Publications

No. 101 **Directory of On-going Research in Cancer Epidemiology 1989/90**
Edited by M. Coleman and J. Wahrendorf
1989; 818 pages; £36.00

No. 102 **Patterns of Cancer in Five Continents**
Edited by S.L. Whelan and D.M. Parkin
1990; 162 pages; £25.00

No. 103 **Evaluating Effectiveness of Primary Prevention of Cancer**
Edited by M. Hakama, V. Beral, J.W. Cullen and D.M. Parkin
1990; 250 pages; £32.00

No. 104 **Complex Mixtures and Cancer Risk**
Edited by H. Vainio, M. Sorsa and A.J. McMichael
1990; 442 pages; £38.00

No. 105 **Relevance to Human Cancer of *N*-Nitroso Compounds, Tobacco Smoke and Mycotoxins**
Edited by I.K. O'Neill, J. Chen and H. Bartsch
1991; 614 pages; £70.00

No. 106 **Atlas of Cancer Incidence in the German Democratic Republic**
Edited by W.H. Mehnert, M. Smans and C.S. Muir
Publ. due 1992; c.328 pages; £42.00

No. 107 **Atlas of Cancer Mortality in the European Economic Community**
Edited by M. Smans, C.S. Muir and P. Boyle
Publ. due 1991; approx. 230 pages; £35.00

No. 108 **Environmental Carcinogens: Methods of Analysis and Exposure Measurement. Volume 11: Polychlorinated Dioxins and Dibenzofurans**
Edited by C. Rappe, H.R. Buser, B. Dodet and I.K. O'Neill
1991; 426 pages; £45.00

No. 109 **Environmental Carcinogens: Methods of Analysis and Exposure Measurement. Volume 12: Indoor Air Contaminants**
Edited by B. Seifert, B. Dodet and I.K. O'Neill
Publ. due 1992; approx. 400 pages

No. 110 **Directory of On-going Research in Cancer Epidemiology 1991**
Edited by M. Coleman and J. Wahrendorf
1991; 753 pages; £38.00

No. 111 **Pathology of Tumours in Laboratory Animals, Second Edition, Volume 2, Tumours of the Mouse**
Edited by V.S. Turusov and U. Mohr
Publ. due 1992; approx. 500 pages

No. 112 **Autopsy in Epidemiology and Medical Research**
Edited by E. Riboli and M. Delendi
1991; 288 pages; £25.00

No. 113 **Laboratory Decontamination and Destruction of Carcinogens in Laboratory Wastes: Some Mycotoxins**
Edited by M. Castegnaro, J. Barek, J.-M. Frémy, M. Lafontaine, M. Miraglia, E.B. Sansone and G.M. Telling
1991; approx. 60 pages; £11.00

No. 114 **Laboratory Decontamination and Destruction of Carcinogens in Laboratory Wastes: Some Polycyclic Heterocyclic Hydrocarbons**
Edited by M. Castegnaro, J. Barek, J. Jacob, U. Kirso, M. Lafontaine, E.B. Sansone, G.M. Telling and T. Vu Duc
1991; approx. 40 pages; £8.00

IARC MONOGRAPHS ON THE EVALUATION OF CARCINOGENIC RISKS TO HUMANS

(Available from booksellers through the network of WHO Sales Agents)

Volume 1 **Some Inorganic Substances, Chlorinated Hydrocarbons, Aromatic Amines, *N*-Nitroso Compounds, and Natural Products**
1972; 184 pages (*out of print*)

Volume 2 **Some Inorganic and Organometallic Compounds**
1973; 181 pages (out of print)

Volume 3 **Certain Polycyclic Aromatic Hydrocarbons and Heterocyclic Compounds**
1973; 271 pages (*out of print*)

Volume 4 **Some Aromatic Amines, Hydrazine and Related Substances, *N*-Nitroso Compounds and Miscellaneous Alkylating Agents**
1974; 286 pages;
Sw. fr. 18.-/US $14.40

Volume 5 **Some Organochlorine Pesticides**
1974; 241 pages (*out of print*)

Volume 6 **Sex Hormones**
1974; 243 pages (*out of print*)

Volume 7 **Some Anti-Thyroid and Related Substances, Nitrofurans and Industrial Chemicals**
1974; 326 pages (*out of print*)

Volume 8 **Some Aromatic Azo Compounds**
1975; 375 pages;
Sw. fr. 36.-/US $28.80

Volume 9 **Some Aziridines, *N*-, *S*- and *O*-Mustards and Selenium**
1975; 268 pages;
Sw.fr. 27.-/US $21.60

Volume 10 **Some Naturally Occurring Substances**
1976; 353 pages (*out of print*)

Volume 11 **Cadmium, Nickel, Some Epoxides, Miscellaneous Industrial Chemicals and General Considerations on Volatile Anaesthetics**
1976; 306 pages (*out of print*)

Volume 12 **Some Carbamates, Thiocarbamates and Carbazides**
1976; 282 pages;
Sw. fr. 34.-/US $27.20

Volume 13 **Some Miscellaneous Pharmaceutical Substances**
1977; 255 pages;
Sw. fr. 30.-/US$ 24.00

Volume 14 **Asbestos**
1977; 106 pages (*out of print*)

Volume 15 **Some Fumigants, The Herbicides 2,4-D and 2,4,5-T, Chlorinated Dibenzodioxins and Miscellaneous Industrial Chemicals**
1977; 354 pages;
Sw. fr. 50.-/US $40.00

Volume 16 **Some Aromatic Amines and Related Nitro Compounds - Hair Dyes, Colouring Agents and Miscellaneous Industrial Chemicals**
1978; 400 pages;
Sw. fr. 50.-/US $40.00

Volume 17 **Some *N*-Nitroso Compounds**
1987; 365 pages;
Sw. fr. 50.-/US $40.00

Volume 18 **Polychlorinated Biphenyls and Polybrominated Biphenyls**
1978; 140 pages;
Sw. fr. 20.-/US $16.00

Volume 19 **Some Monomers, Plastics and Synthetic Elastomers, and Acrolein**
1979; 513 pages;
Sw. fr. 60.-/US $48.00

Volume 20 **Some Halogenated Hydrocarbons**
1979; 609 pages (*out of print*)

Volume 21 **Sex Hormones** (II)
1979; 583 pages;
Sw. fr. 60.-/US $48.00

Volume 22 **Some Non-Nutritive Sweetening Agents**
1980; 208 pages;
Sw. fr. 25.-/US $20.00

Volume 23 **Some Metals and Metallic Compounds**
1980; 438 pages (*out of print*)

Volume 24 **Some Pharmaceutical Drugs**
1980; 337 pages;
Sw. fr. 40.-/US $32.00

Volume 25 **Wood, Leather and Some Associated Industries**
1981; 412 pages;
Sw. fr. 60-/US $48.00

Volume 26 **Some Antineoplastic and Immunosuppressive Agents**
1981; 411 pages;
Sw. fr. 62.-/US $49.60

Volume 27 **Some Aromatic Amines, Anthraquinones and Nitroso Compounds, and Inorganic Fluorides Used in Drinking Water and Dental Preparations**
1982; 341 pages;
Sw. fr. 40.-/US $32.00

Volume 28 **The Rubber Industry**
1982; 486 pages;
Sw. fr. 70.-/US $56.00

Volume 29 **Some Industrial Chemicals and Dyestuffs**
1982; 416 pages;
Sw. fr. 60.-/US $48.00

Volume 30 **Miscellaneous Pesticides**
1983; 424 pages;
Sw. fr. 60.-/US $48.00

Volume 31 **Some Food Additives, Feed Additives and Naturally Occurring Substances**
1983; 314 pages;
Sw. fr. 60-/US $48.00

Volume 32 **Polynuclear Aromatic Compounds, Part 1: Chemical, Environmental and Experimental Data**
1984; 477 pages;
Sw. fr. 60.-/US $48.00

Volume 33 **Polynuclear Aromatic Compounds, Part 2: Carbon Blacks, Mineral Oils and Some Nitroarenes**
1984; 245 pages;
Sw. fr. 50.-/US $40.00

Volume 34 **Polynuclear Aromatic Compounds, Part 3: Industrial Exposures in Aluminium Production, Coal Gasification, Coke Production, and Iron and Steel Founding**
1984; 219 pages;
Sw. fr. 48.-/US $38.40

Volume 35 **Polynuclear Aromatic Compounds, Part 4: Bitumens, Coal-tars and Derived Products, Shale-oils and Soots**
1985; 271 pages;
Sw. fr. 70.-/US $56.00

Volume 36 **Allyl Compounds, Aldehydes, Epoxides and Peroxides**
1985; 369 pages;
Sw. fr. 70.-/US $70.00

Volume 37 **Tobacco Habits Other than Smoking: Betel-quid and Areca-nut Chewing; and some Related Nitrosamines**
1985; 291 pages;
Sw. fr. 70.-/US $56.00

Volume 38 **Tobacco Smoking**
1986; 421 pages;
Sw. fr. 75.-/US $60.00

Volume 39 **Some Chemicals Used in Plastics and Elastomers**
1986; 403 pages;
Sw. fr. 60.-/US $48.00

Volume 40 **Some Naturally Occurring and Synthetic Food Components, Furocoumarins and Ultraviolet Radiation**
1986; 444 pages;
Sw. fr. 65.-/US $52.00

Volume 41 **Some Halogenated Hydrocarbons and Pesticide Exposures**
1986; 434 pages;
Sw. fr. 65.-/US $52.00

Volume 42 **Silica and Some Silicates**
1987; 289 pages;
Sw. fr. 65.-/US $52.00

Volume 43 **Man-Made Mineral Fibres and Radon**
1988; 300 pages;
Sw. fr. 65.-/US $52.00

Volume 44 **Alcohol Drinking**
1988; 416 pages;
Sw. fr. 65.-/US $52.00

Volume 45 **Occupational Exposures in Petroleum Refining; Crude Oil and Major Petroleum Fuels**
1989; 322 pages;
Sw. fr. 65.-/US $52.00

Volume 46 **Diesel and Gasoline Engine Exhausts and Some Nitroarenes**
1989; 458 pages;
Sw. fr. 65.-/US $52.00

Volume 47 **Some Organic Solvents, Resin Monomers and Related Compounds, Pigments and Occupational Exposures in Paint Manufacture and Painting**
1990; 536 pages;
Sw. fr. 85.-/US $68.00

Volume 48 **Some Flame Retardants and Textile Chemicals, and Exposures in the Textile Manufacturing Industry**
1990; 345 pages;
Sw. fr. 65.-/US $52.00

Volume 49 **Chromium, Nickel and Welding**
1990; 677 pages;
Sw. fr. 95.–/US$76.00

Volume 50 **Pharmaceutical Drugs**
1990; 415 pages;
Sw. fr. 65.–/US$52.00

Volume 51 **Coffee, Tea, Mate, Methylxanthines and Methylglyoxal**
1991; 513 pages;
Sw. fr. 80.–/US$64.00

Volume 52 **Chlorinated Drinking-water; Chlorination By-products; Some Other Halogenated Compounds; Cobalt and Cobalt Compounds**
1991; 544 pages;
Sw. fr. 80.–/US$64.00

Supplement No. 1
Chemicals and Industrial Processes Associated with Cancer in Humans (IARC Monographs, Volumes 1 to 20)
1979; 71 pages; (*out of print*)

Supplement No. 2
Long-term and Short-term Screening Assays for Carcinogens: A Critical Appraisal
1980; 426 pages;
Sw. fr. 40.-/US $32.00

Supplement No. 3
Cross Index of Synonyms and Trade Names in Volumes 1 to 26
1982; 199 pages (*out of print*)

Supplement No. 4
Chemicals, Industrial Processes and Industries Associated with Cancer in Humans (IARC Monographs, Volumes 1 to 29)
1982; 292 pages (*out of print*)

Supplement No. 5
Cross Index of Synonyms and Trade Names in Volumes 1 to 36
1985; 259 pages;
Sw. fr. 46.-/US $36.80

Supplement No. 6
Genetic and Related Effects: An Updating of Selected IARC Monographs from Volumes 1 to 42
1987; 729 pages;
Sw. fr. 80.-/US $64.00

Supplement No. 7
Overall Evaluations of Carcinogenicity: An Updating of IARC Monographs Volumes 1-42
1987; 434 pages;
Sw. fr. 65.-/US $52.00

Supplement No. 8
Cross Index of Synonyms and Trade Names in Volumes 1 to 46 of the IARC Monographs
1990; 260 pages;
Sw. fr. 60.-/US $48.00

IARC TECHNICAL REPORTS*

No. 1 **Cancer in Costa Rica**
Edited by R. Sierra, R. Barrantes, G. Muñoz Leiva, D.M. Parkin, C.A. Bieber and N. Muñoz Calero
1988; 124 pages;
Sw. fr. 30.-/US $24.00

No. 2 **SEARCH: A Computer Package to Assist the Statistical Analysis of Case-control Studies**
Edited by G.J. Macfarlane, P. Boyle and P. Maisonneuve (in press)

No. 3 **Cancer Registration in the European Economic Community**
Edited by M.P. Coleman and E. Démaret
1988; 188 pages;
Sw. fr. 30.-/US $24.00

No. 4 **Diet, Hormones and Cancer: Methodological Issues for Prospective Studies**
Edited by E. Riboli and R. Saracci
1988; 156 pages;
Sw. fr. 30.-/US $24.00

No. 5 **Cancer in the Philippines**
Edited by A.V. Laudico, D. Esteban and D.M. Parkin
1989; 186 pages;
Sw. fr. 30.-/US $24.00

No. 6 **La genèse du Centre International de Recherche sur le Cancer**
Par R. Sohier et A.G.B. Sutherland
1990; 104 pages
Sw. fr. 30.-/US $24.00

No. 7 **Epidémiologie du cancer dans les pays de langue latine**
1990; 310 pages
Sw. fr. 30.-/US $24.00

No. 8 **Comparative Study of Anti-smoking Legislation in Countries of the European Economic Community**
Edited by A. Sasco
1990; c. 80 pages
Sw. fr. 30.-/US $24.00
(English and French editions available) (in press)

DIRECTORY OF AGENTS BEING TESTED FOR CARCINOGENICITY (Until Vol. 13 Information Bulletin on the Survey of Chemicals Being Tested for Carcinogenicity)*

No. 8 Edited by M.-J. Ghess, H. Bartsch and L. Tomatis
1979; 604 pages; Sw. fr. 40.-

No. 9 Edited by M.-J. Ghess, J.D. Wilbourn, H. Bartsch and L. Tomatis
1981; 294 pages; Sw. fr. 41.-

No. 10 Edited by M.-J. Ghess, J.D. Wilbourn and H. Bartsch
1982; 362 pages; Sw. fr. 42.-

No. 11 Edited by M.-J. Ghess, J.D. Wilbourn, H. Vainio and H. Bartsch
1984; 362 pages; Sw. fr. 50.-

No. 12 Edited by M.-J. Ghess, J.D. Wilbourn, A. Tossavainen and H. Vainio
1986; 385 pages; Sw. fr. 50.-

No. 13 Edited by M.-J. Ghess, J.D. Wilbourn and A. Aitio 1988; 404 pages; Sw. fr. 43.-

No. 14 Edited by M.-J. Ghess, J.D. Wilbourn and H. Vainio
1990; 370 pages; Sw. fr. 45.-

NON-SERIAL PUBLICATIONS †

Alcool et Cancer
By A. Tuyns (in French only)
1978; 42 pages; Fr. fr. 35.-

Cancer Morbidity and Causes of Death Among Danish Brewery Workers
By O.M. Jensen
1980; 143 pages; Fr. fr. 75.-

Directory of Computer Systems Used in Cancer Registries
By H.R. Menck and D.M. Parkin
1986; 236 pages; Fr. fr. 50.-

* Available from booksellers through the network of WHO sales agents.

†Available directly from IARC

IMPRIMERIE DARANTIERE DIJON-QUETIGNY